Grass Aloes in the South African Veld

by Charles Craib

Illustrations by

Gillian Condy and Murray Ralfe

This is copy number

612

of the STANDARD EDITION numbered 66–2 000

The following editions of this book have been published:

SPONSORS EDITION numbered 1–10

COLLECTORS EDITION numbered 11–60

PRESENTATION COPIES numbered 61–65

2005

Umdaus Press

Grass Aloes

in the South African Veld

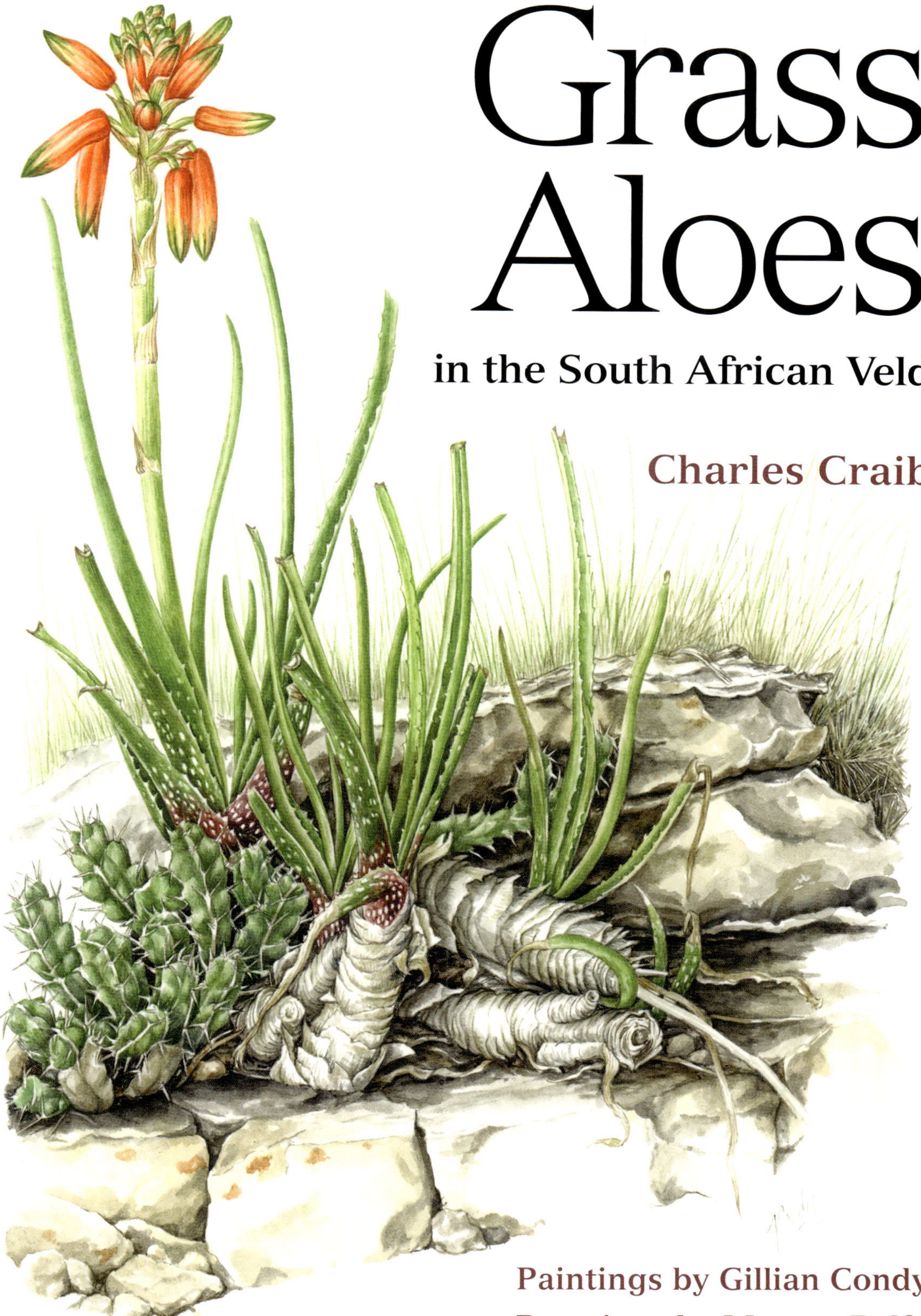

Charles Craib

Paintings by Gillian Condy
Drawings by Murray Ralfe

UMDAUS PRESS
P.O. BOX 11059
HATFIELD 0028
SOUTH AFRICA

E-mail: umdaus@succulents.net
Web site: www.succulents.net

FIRST PUBLISHED 2005

PRODUCTION
Alex Fick and Kotie Retief

EDITING
Emsie du Plessis

DESIGN
Tersia van Rensen

COLOUR SEPARATIONS
Jason Pyper

BINDING OF SPECIAL EDITIONS
Peter Carstens

PRINTING
Tien Wah Press, Singapore

ISBN 1-919766-39-1 (Sponsors Edition)
ISBN 1-919766-40-5 (Collectors Edition)
ISBN 1-919766-41-3 (Standard Edition)

I am delighted to offer these words of congratulation and encouragement to the author, artists and publisher of this wonderful book on the "Grass Aloes in the South African Veld." They should be justly proud of their achievements.

The author, Charles Craib, who has done an incredible amount for the world of botany, particularly in the fields of succulents and bulbs, has again excelled with this magnificent publication. Charles' last book "Geophytic Pelargoniums", with Gillian Condy as the artist and Umdaus Press as Publisher, was out of print within a very short time – proof of his popularity as an author of botanical works.

This book is a wonderful combination of the academic study of the South African Grass Aloe habitat and the amazing artwork of two local artists. Charles' splendid job in assembling the artists and publisher, Umdaus Press, is once again to be applauded.

Gillian Condy, whose work I have much admired for many years has excelled with the exceptionally beautiful watercolours she so meticulously executed for this book. Forty three glorious watercolours which are scientifically accurate, depict perfectly the plants in their micro-environments.

Coupled with Gillian's work are the truly amazing, accurate and detailed pencil drawings by Murray Ralfe, depicting the habitat in which these plants grow.

The descriptions given by Gillian and Murray of the creation of their art-works for this book are a truly innovative exercise and I commend Charles' vision in including these.

The footprint of the development of mankind across our planet is pervasive and generally indiscriminate. Southern Africa is no exception in terms of the destruction of the ecology and habitat of so many plant species by human interference.

This book will assist significantly in preserving our legacy of grass aloes for future generations. Some of these plants, as Charles points out, will surely be lost forever due to habitat degradation. I wholeheartedly support this book and those to come. May there be many more to add to the invaluable archive of Botany for future generations to behold.

Steve Bales

Group Art Consultant
FirstRand Banking Group
Sandton

November 2005

This book is dedicated to the memory of the late

Gilbert Westacott Reynolds,

highly accomplished authority on the genus Aloe

The G.W. Reynolds Gate at the Pretoria National Botanical Garden.

The gate commemorates the life of Gilbert Westacott Reynolds (1895–1967) and this drawing symbolises the entrance to this book and the enormous and unsurpassed contribution Reynolds made to the study of the genus *Aloe*.

The gate was made by craftsman metalworker Hans Brugger (1919–1982) and is amongst the largest and most impressive of his works. It is one of two gates featuring indigenous flora at the Pretoria National Botanical Garden. The Reynolds Gate, with its robust aloes, has a strongly sculptural component whereas the other gate, designed for the official opening of the Botanical Garden in July 1973, features the intricate designs of several indigenous grasses.

Contents

Foreword viii
Preface x
Acknowledgements xi
Introduction 1
Painting the plants 2
Drawing the landscapes and habitats 7
Taxonomy and the sociology of botanical knowledge 10
Afforestation and the spread of alien plants in grassland 13
The changing landscape: mining, urban development and informal settlements 17
Fire and burning in the habitat 20
Habitats and growth cycles 23
Pollination 25
Development and dispersal of seed 27
Road and railway-line reserves 28
The plants 30
1. *Aloe albida* 30
2. *Aloe boylei* 33
3. *Aloe chortolirioides* 36
4. *Aloe cooperi* 39
5. *Aloe craibii* 44
6. *Aloe dominella* 47
7. *Aloe ecklonis* 50
8. *Aloe fouriei* 53
9. *Aloe hlangapies* 57
10. *Aloe inconspicua* 62
11. *Aloe integra* 66
12. *Aloe kniphofioides* 71
13. *Aloe kraussii* 75
14. *Aloe linearifolia* 78
15. *Aloe micracantha* 81
16. *Aloe minima* var. *minima* 85
17. *Aloe minima* var. *blyderivierensis* 88
18. *Aloe modesta* 91
19. *Aloe myriacantha* 94
20. *Aloe nubigena* 98
21. *Aloe parviflora* 101
22. *Aloe saundersiae* 104
23. *Aloe soutpansbergensis* 107
24. *Aloe thompsoniae* 111
25. *Aloe verecunda* 114
26. *Aloe vossii* 119
27. *Aloe woolliana* 123
28. *Chortolirion angolense* 127
Cultivation 137
Grass aloes in a veld garden 141
Survival in a changed environment 142
References 144
Glossary 145
Index to scientific names of organisms 149
List of subscribers 151

Foreword

The fascinating world of the grass aloes of South Africa

The history of the grass aloes dates back to the early 1800s when two collections of species new to science and horticulture from the Eastern Cape, *Aloe micracantha* and *Aloe myriacantha*, were introduced to Europe in quick succession through the collecting efforts of James Bowie (ca. 1789–2 July 1869). Perhaps not surprisingly, the earliest description of the first species to have been published was that of *A. micracantha*, the South African grass aloe with the furthest known westerly distribution, typically in grassy fynbos. Between the time that Bowie, who worked in the Cape of Good Hope as a professional collector of plant specimens for the Royal Botanic Gardens Kew, United Kingdom, collected these plants and the benchmark works of Gilbert Reynolds on aloes which appeared about 150 years later, numerous new grass aloe species were described, virtually all of them small, low-growing plants with slender, grass-like leaves and robust underground perennating structures.

Above: The Pinnacle immediately north of Graskop in Mpumalanga. These steep escarpment cliffs are the habitat of *Aloe nubigena*. This area is subject to frequent and prolonged misty weather in summer. The aloes typically grow amongst moss and grass near the summits of the cliffs.

Among their relatives in the genus *Aloe*, the grass aloes play an unfamiliar, muted role. For this there are undoubtedly a number of reasons, not the least of which being that some species are not easy to keep in cultivation, and of course the fact that they are generally small in size and stature and in most cases not easy to detect in their natural habitat. Many of their larger kin can be easily observed, and even identified, from a vehicle travelling along a highway at great speed. Therefore, what Charles Craib is perhaps really doing most remarkably in this book is rescuing grass aloes from the dreary bondage of obscurity. And, really, as Charles shows us here, there should be no reason why this should have been necessary as grass aloes have exquisitely beautiful flowers that certainly warrant the effort to keep them flourishing and flowering. To grow them well simply comes as an additional challenge to keep these fascinating plants thriving in artificial landscapes and in specialist collections.

Gathering and compiling the information and then analysing it to write a tome on any group of plants, is not a simple matter. When going to the level of detail that Charles has, it is clearly a labour of love, and one requiring an almost superhuman effort. Indeed, if one was not confident enough to think it possible to create sense—a framework essentially—from the nearly impossible mass of apparently random characters that make up a species, if one could not imagine oneself conducting a range of explorations and investigations, then one could not possibly attempt to unravel the intricacies of grass aloe taxonomy. But, much more than that, in this book Charles shows us the many worlds of grass aloes, from the Eastern Cape coast to the high-altitude mountains of Mpumalanga, and beyond.

Over the past several years I have often been confronted with specimens—both dried and living—and information on purported new species of *Aloe*. Eight out of ten times it was a strange or somewhat aberrant form of a species of grass aloe. If this is an indication of the popularity of this fairly well circumscribed infrageneric group of *Aloe*, then it must rate as one of the most widely cultivated groups in this horticulturally attractive genus. (Incidentally, the other two times, I was confronted with an aberrant species of maculate *Aloe*.) Sadly, the number of aberrant forms of grass aloes that I have seen over the past several years does not reflect in the popularity of this group. Given that the flowers of grass aloes are often large, brightly coloured and carried in short but dense inflorescences, any gardener should be easily persuaded to give them a try. And this is another benefit of Charles's approach to this topic: the book carries a wealth of information for the uninitiated who would like to learn more about growing representatives of this group of aloes. Given his keen sense of observation, it is not surprising that the only new species of grass aloe came from one of his own collections, and I had the privilege of describing it and naming it after him, as *Aloe craibii*.

But all is not well in the natural habitats of grass aloes. As the ruthless hand of humans today stretches almost uncompromisingly into the far corners of our wonderful country, nature lovers should increasingly be made aware of the dangers of over-exploiting these species. Perhaps the greatest danger to the survival of natural populations of grass aloes is that they ever so often go undetected in their habitats. This is quite predictable, as they are not called grass aloes for nothing: their leaves closely resemble those of grasses. For example, in some areas where they occur on the boundaries of urban areas, natural veld is mowed by industrial lawnmowers to keep the grass short and to limit the presence of undesired rodents. A further example of a danger to their survival is that grazing for livestock is burned regularly in the dry winter months of the South African interior, to 'prepare' these pastures for grazing. And of course these areas are grazed and trampled by animals shortly thereafter! On the other hand, the difficulty with which grass aloes are detected in their habitat also comes as a protective measure, as one usually has to search quite hard to find these plants in the field.

This book is unquestionably contributing to the dissemination of information on the grass aloes, especially since the text was done with such gusto and virtuosity. Charles is essentially contextualising grass aloe biology and natural history in the web of available *Aloe* information. However, he will be the first to admit that his work on the grass aloes is a beginning, not an end. The curious and observant *Aloe* scholar is undeniably invited to enquire, even to challenge and reassess the interpretation at hand.

But this work is not only rich in informative text, it is also exquisitely illustrated with scientifically accurate botanical paintings by Gillian Condy. Along with the other illustrative material of Murray Ralfe, the book carries a wealth of information and illustrations of these plants and their natural habitats. On this score, all who are interested in the rich South African succulent flora, particularly the aloes, should acquire a copy of this book. Like the earlier work of Charles Craib on the geophytic pelargoniums (Umdaus Press, 2001), it will undoubtedly become a much sought-after resource and instant Africana.

Charles had the dedication required for the task of popularising the grass aloes. This work will stand as a milestone in the field of research on these plants.

Gideon F. Smith, Ph.D. (Pretoria)

Chief Director: Research and Scientific Services,
South African National Biodiversity Institute,
Private Bag X101,
Pretoria, 0001
South Africa

and

John Acocks Professor of Botany,
H.G.W.J. Schweickerdt Herbarium,
Department of Botany,
University of Pretoria,
Pretoria, 0002
South Africa

Preface

The grass aloes are a group of *Aloe* species characterised by foliage that in habitat dies back fully or partially in the winter months. The group ranges in size from very small species such as *Aloe albida* and *Aloe saundersiae* to some robust plants with broad, conspicuous leaves like *Aloe boylei* and *Aloe integra*.

Grass aloes grow mostly in grassland, either flat, hilly or mountainous. Three species, *Aloe nubigena, Aloe soutpansbergensis* and *Aloe thompsoniae*, grow on cliff ledges amongst moss and grass. These three are typical cliff dwellers in much of their range but also occur amongst boulders on steep, rocky mountain tops.

South Africa's grass aloes are all found in the eastern summer-rainfall parts of the country. The most southerly distribution of the group is the Humansdorp and Kareedou areas of the Eastern Cape. Their northernmost distribution in the country is the Soutpansberg. This mountain range provides the habitat for two endemic species.

Grass aloes are either succulent shrublets with several stems from which annual leaves appear, or stemless plants with leaves arising directly from the underground parts. The latter habit characterises some of the smallest species of grass aloes. Two species, *Aloe kniphofioides* and *Aloe modesta*, have bulb-like swellings.

One new grass aloe species has been described in recent times, namely *Aloe craibii*. The plants appear to be confined to a few sites in the mountains near Barberton and may also be found in similar habitat in the adjacent part of Swaziland.

Grass aloes may be subject to taxonomic revision as techniques in the knowledge production process are changed, improved and refined. This process may lead to the grouping of some species under one name or to the revision of genus and species concepts altogether. The names used in this book are those used by Reynolds (1969: 110–155).

The book consists of a detailed account of the grass aloe species in habitat and in cultivation and represents the culmination of some 20 years of active research on this group. An account is also provided for *Chortolirion angolense*. The growth cycle and habit of this monotypic genus have much in common with those of many grass aloes and it was studied at the same time as the data for this book were collected.

All the water colour paintings of the different grass aloes and the pencil drawings of their habitats have been prepared *in situ*. This has entailed much travelling throughout eastern South Africa where the different species grow.

St Cyprian's Church at Highlands near Grahamstown in the Eastern Cape. It was originally the engineer's residence during construction of the Alicedale-Grahamstown railway line around the late 1870s but was converted into a church after the railway line had been completed. The building was officially dedicated as St Cyprian's Church on 29 November 1893. The railway-line reserve between Coldspring and Highlands, southwest of Grahamstown, is most significant. It served the purpose of protecting *Aloe myriacantha* after sheep and cattle farming was introduced in the grasslands of the Albany District after the arrival of the 1820 Settlers.

Acknowledgements

The publication of a book such as this is not possible without the input of many people with a wide range of skills and talents. I would like to thank Braam van Wyk, Professor of Botany at the University of Pretoria, for reading and commenting on an earlier draft of the text. Braam is particularly interested in and knowledgeable about grassland flora and grassland ecology. Thanks are also due to Emsie du Plessis for meticulously editing the text.

Gideon Smith, Chief Director: Research and Scientific Services at the South African National Biodiversity Institute, has kindly written the Foreword. I would like to thank him for this and also for his continued interest in and encouragement with this project.

Welland Cowley of Port Elizabeth and Philip Crous and Charl Malan of Grahamstown were very helpful with the field research on *Aloe micracantha* and *Aloe myriacantha*. They also supplied important information on the propagation and cultivation of these plants.

Bernard de Souza, formerly of Barberton in Mpumalanga, provided interesting information about a possibly undescribed grass aloe in the mountains near Barberton. He is also thanked for arranging a field trip to search for these plants.

Neil Crouch of Durban is thanked for assistance with the field research on *Aloe linearifolia*. He also kindly provided Gillian Condy with flowering material for completion of the habitat painting of *Aloe parviflora*.

Roger and Wilna Stones went to much trouble to organise a field expedition to the summit of the Wolkberg in Limpopo Province. I conducted research on *Aloe thompsoniae* during the course of this visit and Gillian Condy painted plants in flower in their mountain-top habitat. Maddy Lehman of Johannesburg kindly supplied the transport for Gillian and me. She also contributed to discussions about the autecology of the plants during the course of the fieldwork.

Fanie Venter formerly of Polokwane (now residing in New Zealand) and Andrew Hankey of the Walter Sisulu National Botanical Garden in Gauteng deserve special thanks. They arranged for fieldwork in Limpopo Province to study *Aloe fouriei* growing on the Highlands Mountains and also on the summit of the Strydpoort Mountains.

Andrew Hankey took part in numerous valuable discussions about grass aloes in habitat and cultivation and contributed significantly to the research on *Chortolirion angolense*. He also contributed a great deal to discussions about the *ex situ* conservation of grass aloes.

The success of the research on *Aloe vossii* and *Aloe soutpansbergensis* is attributable to the significant input of Norbert Hahn. Norbert provided valuable habitat information about the two species and accompanied Gillian Condy and me in the field.

Jürgen Greve of Hermannsburg near Greytown kindly facilitated research on *Aloe boylei* and *Aloe kraussii*.

David Green of Estcourt in KwaZulu-Natal helped considerably with field research on *Aloe inconspicua*. I extend my sincere thanks to him for this. He is also thanked for lending me copies of correspondence between himself and various botanists, which eventually led to the formal description of this species.

John Lavranos, resident in Portugal, assisted greatly with discussions about the autecology and taxonomy of grass aloes during the course of this research. He also assisted with locality data at various times during the course of this project.

I would particularly like to thank the group of people to whom I refer as 'the production team'. Gillian Condy and Murray Ralfe, the two artists, went to a great deal of trouble to ensure that their artwork was of the highest standard. They were also most helpful in completing significant artwork at short notice. Tersia van Rensen went to a great deal of trouble to ensure the best possible layout of the text and illustrations. Geraldine Dittrich typed the manuscript from a draft hand-written in pencil. I am most grateful to her for this, particularly for incorporating the numerous additions and alterations that I made while self-editing my text, prior to the formal editing process.

The directors of Umdaus Press went to great lengths to meet the exacting requirements for the publication of this book. I am very grateful to them for this and particularly for permitting the two artists and myself to make a significant contribution to the design and layout process.

Carol Knoll and Connall Oosterbrook, both of Johannesburg, deserve special thanks for photographing various grass aloe species in habitat. Their photographs form an important record, particularly where pollinators of various grass aloes have been photographed on the flowers. They also kindly photographed the author and Gillian Condy at work in the field—these photographs have been used for the dust jacket.

Mike Picker of the Zoology Department of the University of Cape Town helped greatly with the identification and natural history of grass aloe seed parasites. This was most helpful and is much appreciated.

Tamantha Smith of the South African National Biodiversity Institute facilitated the identification of various grass aloe pollinators and seed parasites. She is thanked for these important contributions.

John and Sandie Burrows of the Buffelskloof Nature Reserve helped with locality data for *Aloe woolliana* and offered hospitality to Gillian Condy and Murray Ralfe. This is much appreciated.

Klaas Devenish, who farms near Wakkerstroom, supplied much significant information on *Aloe modesta*. He is sincerely thanked for his input.

Pieter Winter of the South African National Biodiversity Institute in Pretoria contributed significantly to discussions about the taxonomy of grass aloes. His taxonomic acumen concerning this group of plants is greatly valued and his input much appreciated.

Joan Symons kindly supplied seeds of *Aloe inconspicua* which were used for research purposes. She is warmly thanked for this important contribution.

Finally I would like to thank my friend Joan de Villiers who frequently accompanied me on field trips to do research, sometimes in very misty inclement weather.

Charles Craib
Johannesburg, June 2005

The habitat of *Aloe cooperi* high up on the summit of the southern Leolo Mountains in Sekhukhuneland where it grows amongst tumbled boulders. It more rarely occurs in short open grassland, often trampled by cattle, sheep and goats kept by the pastoralists who live on the mountain.

Introduction

The grasslands of eastern South Africa are undergoing a great deal of change. Much of this has been induced by alterations in land-use patterns and by the expanding human population.

Vast amounts of land in Limpopo Province, Mpumalanga and KwaZulu-Natal have been planted with timber, destroying many grass aloe populations in the process. The process of afforestation has had the effect of restricting various grass aloe species to small tracts of land between alien timber plantations. Most of these populations are not viable in the long term and only a few are showing active regeneration.

Railway-line reserves and road reserves have played a particularly significant role in preserving limited numbers of several grass aloe species. These areas are not subject to the same range of pressures encountered in most grassland. They are not grazed by domestic stock and seeds that fall from the parent plants stand a good chance of germination. Some large groups of various grass aloes are found in these reserves. There is a growing tendency nowadays, particularly around informal settlements, for wire fencing to be removed by local people. The people in the informal settlements are poor and often unemployed and need to make use of whatever building and fencing materials are at hand. The railway-line and road reserve populations of grass aloe species can be expected to undergo a decline in many areas on the outskirts of towns. In places where this occurs, the strong populations in road and railway-line reserves will no longer be available to repopulate the surrounding veld.

Mining plays a significant role in the destruction of some grass aloe habitats and in the preservation of others. Open-cast or strip mining destroys grass aloe habitats but this is overall of little significance to grass aloes at present. In most instances open-cast mining takes place where relatively common grass aloes are found. In western Gauteng, deep-level gold mining has played a significant role in preserving habitats. Mining activity has kept out other forms of land use such as small-scale stock farming with its associated trampling of the habitat.

These processes are discussed in some depth under the species accounts in the text and in the chapters on grass aloe habitats.

As the predicted consequences of global warming materialise, the populations of most grass aloes can be expected to shrink. This may also lead to the extinction of several of the dwarf species, all of which require a well-watered, moist habitat during their growing and flowering season. Most grass aloe species are integrally linked to periodic grass fires. Should there be an increase in savanna and scrub, precipitated by global warming, it is likely that many of the species will not be able to flower and seed adequately. The consequences of global warming with respect to grass aloe populations are speculative at this stage. There is, however, the example of *Aloe fouriei*, a species of mostly rather arid dolomite hills and mountains. The effects of prolonged drought, over-grazing and habitat alteration are documented in the species account for this plant. Preliminary indications are that regeneration from seed hardly exists in some populations and that, over time, adult plants will gradually disappear. It is true to say that our planet has warmed and cooled many times and that there are no data to indicate how grass aloe species fared during these periods. They are, however, the one group of *Aloe* in South Africa that are particularly vulnerable to global warming, habitat alteration and habitat degradation.

The intricate relationship all grass aloes have with their environment is documented for the various species. This serves several purposes concerning the supply of information. It records the factors that lead to the decline of grass aloe numbers and in a few cases, to their increase. A foundation is also laid for the stimulation of debate as to appropriate methods for conserving various species. A significant role of books such as this one is to provide a historical account of the factors that lead to plant extinctions, in this case the likely extinction of most of South Africa's grass aloes.

Fire plays a critical role in the health of all grass aloe populations, except possibly *Aloe nubigena* which is almost entirely a cliff dweller. As it is difficult to replicate these conditions in cultivation, several species are not doing well in horticulture.

Some grass aloe species produce enormous amounts of seed over a limited period of a few weeks in late summer and autumn. Others are shy seed bearers but their clump-forming habit and extended flowering period ensure that some seed is set free into the environment over several months. The seeds of grass aloes must last for at least two seasons to ensure regeneration. Germination is usually at its best the season after it has been set, but some may germinate if there are good autumn rains the year the seeds were produced. The seeding period for all grass aloes occurring in South Africa is from December to April. The first ones to set seed are the early summer-flowering species such as *Aloe kniphofioides*.

The amount of seed set varies hugely from one season to the next in the case of particular species. Grass aloe seeds are sometimes very heavily parasitised and are also subject to fungal attack during prolonged wet weather.

Much of the viable seed successfully liberated from the parent plants is destroyed by winter grass fires. However, in the case of heavy autumn rains in April, the runoff tends to bury much of the wind-blown seed at the base of grass tussocks where they accumulate and are somewhat protected against the heat. The grass fires have an interesting double role in the autecology of grass aloes. They create ideal conditions for germination and regeneration of most grass aloe species and also burn quantities of seeds that have not already lodged in the soil.

Several grass aloe species that are prolific seeders have become quite numerous in areas where soil has been artificially disturbed. This has generally occurred after heavy grazing and as a consequence of the building of roads and railway lines. Attention is drawn to this process in the species accounts of some of the larger, robust grass aloes.

An account of *Chortolirion angolense* is also provided in this book. Although this plant is not a grass aloe, its autecology is very similar to that of many grass aloe species.

Painting the plants

Two principal types of art work were selected for this book—watercolour paintings of all the grass aloe species in their immediate habitat, and pencil drawings of various aspects of the countryside in which the plants are found. Botanical artist Gillian Condy was involved from the onset of the project and was responsible for the watercolour paintings.

Grass aloes are a group of plants that lend themselves to fine botanical illustration, making it easier for the artist to portray them correctly. In cultivation, the plants are rarely grown in their natural form, thus the only method of depicting them true to life, is to paint them in their habitat. This decision, taken jointly by the author and artist, resulted in numerous trips throughout eastern South Africa. The paintings of the aloes in habitat are as significant as the text itself, being a visual record of many aspects of the annual growth cycle of the plants. The paintings were executed over a period of some ten years, during which time the author observed the populations in varying seasons.

In most cases, the author located several populations of the plants in various kinds of habitats before the artist went to a site. The localities most suitable for conveying the information discussed in the book were selected. Wherever possible, the author went to check on the state of flowering of each species in advance of the painting trip. This did not always work out quite as planned. Owing to different weather patterns, some visits coincided with the end of the peak flowering season and the plants were then depicted in their seeding habits, followed by flowers in the subsequent season.

Each grass aloe species grows in a unique microhabitat packed with information about the autecology of the plants. In addition, the greater habitat of grassland, cliff faces or rocky mountain summits are also full of information to interpret and record. In the hope of achieving an authentic and visually pleasing record, the artist decided that each species should be painted in its immediate surroundings. Based on this decision, an accurate and aesthetically pleasing style was developed.

One of the most significant aspects of the painting of each species has been the choice of the subject for illustration. Arriving at a chosen habitat, the author and artist walked around extensively, looking for plants that were typical and representative of the growth, habit and form of the species in question. The tendency was to choose plants growing in an aesthetically pleasing location, where the microhabitat was not too concealed by clumps of grass.

The plants were always drawn in great detail in pencil directly on to the watercolour paper, the artist sitting at ground level or close to it. The distance between the artist and the subject varied according to the scale of the species and the part of the plant or its microhabitat that was being drawn. It was only possible to work in this position for short spells without stretching the limbs. The intense heat and the glare of the paper in harsh sunlight made it very tiring to work out of doors. Luckily the drawings were usually completed in about three hours, the maximum time it proved possible to work in the field.

Colour was impossible to add in the field since it dried on the brush before it could be applied to the paper. The artist kept accurate colour notes on the sheet and worked from flower and leaf specimens under studio conditions once a field trip had been completed. Photographs, which the artist took in the field, were used as an aid to painting the habitat, along with small rocks, grass tufts and soil which had been collected. Owing to time constraints and the remoteness of many of the habitats, it was necessary to complete the initial drawings before leaving.

Arches 300 g/m^2 hotpress watercolour paper was used. The artist preferred not to stretch the paper, so as not to disturb the paper sizing. Many of the subjects were on the small side, not requiring large areas of wash, but smaller brush strokes. If necessary, a painting can be stretched once completed, by lightly spraying the reverse with water and stretching the paper as normal. The artist drew directly on to the watercolour paper with a 2H pencil, with the minimum amount of erasing.

To keep the paper firmly attached to the drawing board, bulldog clips were used, which also held in place wide strips of acetate. They covered areas not being worked on to protect them from insects, birds or the frequent dust storms encountered in the field. The artist rested her hand on a wide piece of cotton fabric, to absorb any moisture. A useful addition to the field bag was found to be hand wipes or a moist facecloth to keep the hands clean.

Since her early student days, the artist has used Winsor and Newton artist's watercolours. These days she tries to keep to a limited range of six colour pigments (French Ultramarine, Cerulean Blue, Cadmium Yellow, Cadmium Lemon, Permanent Rose and Scarlet Lake), with only a few other favourite colours for the occasional finishing touches. She only uses Kolinsky sable watercolour brushes as, with proper care, they hold their shape and spring so superbly.

It was both fortunate and fortuitous that during the estimated 40 field days of painting the aloes, very few excursions had to be abandoned as a result of rain. On one occasion, thick mist and rain made it impossible to paint *Aloe nubigena* near Graskop. Plans were rescheduled and *A. fouriei* was painted instead. The *A. fouriei* locality was only about 80 km away, but in a more arid, rain-shadow area where it was only cloudy and not raining.

The return field trip to *Aloe nubigena* did not offer the most comfortable of localities in which to work. The plants grew on the vertical rock faces of the escarpment and the artist had to manoeuvre along the cliff top and 'perch' on the edge, with a frightening drop inches away. Even then she had to lean over the edge, during gusting winds, to get a reasonable view of the chosen plants. The artist does not enjoy heights. The first locality offered for painting *A. soutpansbergensis* would have involved the artist suspended halfway up a cliff on a chain ladder. This was politely declined, and specimens on top of the cliff were drawn instead!

During the first visit to Piet Retief to illustrate *Aloe hlangapies*, the artist was left working while the author drove off to look at plants in a few other localities. In the late afternoon, a black cloud moved in rapidly and proceeded to empty its contents, accompanied by threatening

The detailed pencil drawing of *Aloe woolliana*. A beautiful rock setting was chosen. The final drawing of the flowerhead and the colour were added in the studio. Approximately three hours were spent in the field, and two hours in the studio.

The painting in progress. During another four and a half hours in the studio, the painting was worked up, along with features in the immediate vicinity. The artist worked with a strong light source from the right, simulating the sun, to give a good three dimensional feel to the painting.

The completed artwork. Photographs of the same plant and a strong mental image of it, together with collected rock samples, were used to complete the painting. The artist has tried to limit the layers of watercolour to a maximum of three or four washes to retain a freshness and the translucent quality of the watercolours. Another five hours were spent to complete the painting.

lightening. Luckily it was one of the few field trips when the artist had taken an old beach umbrella. The artist with the artwork took refuge beneath the closed arms of the umbrella and patiently waited until the author returned. Mpumalanga and the Highveld are renowned for violent summer afternoon thunderstorms.

Several other hardships were encountered on the painting field trips. Grassland absorbs much of the sun's heat. Sitting at ground level, or just off it, the artist had to contend with ground heat as well as the intense, direct sunlight, numerous insect bites, and she was thankful, however, that only one snake was encountered.

Wind of some kind was usually present. While drawing a specimen of *Aloe dominella*, with about 30 flowering stems, growing near Rorke's Drift in KwaZulu-Natal, a strong gust of wind pulled the drawing board from the artist's hands and ripped the paper. Fortunately the tear was just outside the image area. Because of the field conditions, much reconstruction work had to be done in the studio.

An equally challenging painting was that of *Aloe integra* in dormancy, with a mass of dead leaves, in layers, among dry grass and all with a very similar, soft brown colour. This painting appears on page 69. It provides the basis for an important discussion about *Aloe integra* in its dormant state in the chapter on that species as well as about the relationship of the dry leaves to burning in winter grass fires.

Cold winter weather reduced the number of painting hours that could be endured when out in the field. Near Carletonville, working on the second *Aloe verecunda* plate, there was no shelter from the icy wind. It became so cold that the artist was unable to hold her pencil and the illustration was completed later.

It could be said that painting all the South African grass aloes in habitat was an ambitious undertaking. Talent, experience and professionalism permitted the artist to overcome all the many challenges. The beauty of the different grass aloe habitats and their spiritual nature helped to drive the process. A very special bond can develop between an artist and subject as they experience the same sun and wind. The artist said she often felt she was intruding on the scene when she first arrived at a site. But, the longer she sat quietly, focused on the shapes, colours, textures, changing light and insect activity in the immediate surroundings, the more she felt totally at ease, almost as though under a spell.

An account of the techniques used for drawing and then painting *Aloe woolliana* is given on pages 3–5. The same methods were used for all the paintings in this book. It is hoped that readers enjoy the paintings for the information they convey as well as for their visual beauty.

Drawing the landscapes and habitats

During the early phases of the project it was decided that pencil would be the ideal medium for the landscape illustrations. The challenge was to find a pencil artist accomplished in the field of landscape drawings. Artists skilled in the medium of pencil illustration are rare and at times it seemed as if the idea of including pencil landscape drawings would not be realised.

Late in 2003 the author met Murray Ralfe while he was adding the final details to a drawing of a baobab tree. It was immediately realised that he would be the right artist for the landscape illustrations featured in this book. A meeting was arranged between Gillian Condy, Murray and the author and shortly afterwards it was decided to proceed with the pencil drawings.

The author made a detailed list of the habitats where landscape illustrations were required; these localities were in the Eastern Cape, Gauteng, KwaZulu-Natal, Limpopo Province and Mpumalanga. The artist then made his choices from the guidelines supplied by the author. This latitude provided for interpretation, which was considered important since certain perspectives in some landscapes are more suitable for illustration than others. Pencil is a sensitive medium for conveying information about habitats in which grass aloes are found; the drawings bring to life the habitats occupied by the different species.

A representative selection of grass aloe habitats was drawn, as well as some interesting and significant smaller features in specific landscapes. These, for example, feature the spectacular *Zantedeschia jucunda* growing on the Leolo Mountains in Sekhukhuneland. This arum lilly shares its rocky habitat on the summit of the Leolo Mountains with *Aloe cooperi*.

Murray and Gillian have combined their talents to produce two illustrations. One is of *Aloe hlangapies* in its boulder-strewn hillside habitat near Piet Retief, the other depicts *A. woolliana* growing on an escarpment cliff face near Lydenburg, both in Mpumalanga. First Gillian painted the grass aloe in the foreground, then Murray drew the habitat in pencil as a background. This technique, shared by two artists working on the same sheet of paper, is probably unique in botanical illustration.

The paragraphs that follow, contain Murray's account of his techniques for preparing the landscape and habitat drawings in the field and in the studio. Also included is a discussion of some of the specific illustrations that feature in the book, highlighting certain techniques suited to drawing rocks, grasses and trees in the different landscapes.

The western Soutpansberg drawn from the bushveld below the southern aspect of the mountain range. Letjuma, the highest peak in the Soutpansberg, at an altitude of 1 747 metres, is visible in the far distance in the centre.

The techniques

Each landscape where grass aloes were found, was handled in a similar manner. During the course of a field visit, I walked around the habitat until I found the composition I was looking for. The next step was to draw a few rough sketches of different elements in the composition. I looked for features in the landscape that created curves and strong lines to balance the drawing. This ensured that the illustration contained aesthetic symmetry, the cornerstone of any successful composition.

The landscape I selected was divided into three shapes, namely foreground, middle ground and sky. I looked for the manner in which these shapes interacted with, complemented, or created a tension with each other. After this, I examined the details of the landscape that were more dominant and less dominant. These were combined with shapes in the landscape, for example a dark timber plantation on one side had to be balanced with another feature in the landscape that was also dark, or else a white space.

When I found a composition that appealed to me, I used a digital camera to photograph the site as my eye saw it. To achieve this, I adjusted the zoom lens until it matched my eyesight. Afterwards I took a series of portrait photographs, overlaying each one by about a third. The photographs must be taken in portrait so as to get the least lens distortion. At some sites I took more photographs than at others, but the average number taken per drawing was 15. Afterwards I took photographs of a second habitat perspective for a comparison. The final stage, during the field visit, involved an investigation of features in the landscape I intended to draw. I looked at the grasses and other plants to gain an understanding of the site. I also had to see how grasses and rocks featured in the overall landscape. In addition, I had to make some emotional contact with the landscape.

In the studio at home, I first downloaded the photographs from the digital camera, and categorised them according to the different panoramas found in the landscape. I also identified various plants that featured in the landscape photographs. Then, using my rough composition sketch as a reference, I joined the photographs together, overlaying them to create a seamless image of the site.

The only paper I use for drawings, is Schoellerhammer paper. I have tried others, but cannot achieve as good an effect. Schoellerhammer paper is more effective for my purposes than Schoellerhammer board.

Hard pencils I find the best, from 9H upwards and ending with a 2H or HB. I have tried various makes of pencils, with the Derwent and high-grade Staedtler Mars Lumograph the best.

I drew with the composite image of the site in front of me on my computer screen. The images were large, with the result that I could zoom into detail on the screen when required, while I was drawing. When I had to look at a landscape as a whole, I simply zoomed out. If I had to see detail in dark shadows, I lightened the image to reveal these details. For me, one of the keys to a successful drawing is the balance of light, and revealing a bit of what lies in the shadows, gives depth. I worked on a couple of drawings simultaneously to keep the images of the different landscapes fresh in my mind's eye. This method also has the advantage of preventing my getting embedded in details and thereby losing the essence of a particular drawing.

Some drawings took much longer to complete than others. This is not easy to explain, but some flowed smoothly while others dragged on. If I identified with various shapes in a landscape, the drawings went smoothly. For example, the drawing of St Cyprian's Church went well because of the geometric form of the building. I like geometric shapes. The drawing of the *Aloe nubigena* habitat near Graskop took longer. The indigenous forest in the gorge below the sheer cliffs was difficult to capture. In addition, the drawing was one of the largest formats used in my illustrations. The large drawings took a long time to complete, while those where a smaller format was used, were well suited to my drawing techniques. Ideally, I like to capture the essence of something with as few pencil strokes as possible. This keeps the images clean, crisp and fresh.

The drawings all required a good deal of thought as well as the development of specific techniques. A special method had to be developed for drawing clumps of long grass at Greenhills east of Grahamstown, the habitat of *Aloe micracantha*. The *A. linearifolia* habitat on the outskirts of Margate took a long time to draw, while the *A. micracantha* landscape, consisting of flat coastal grassland near Port Elizabeth, took less time. The illustration featuring the railway-line reserve habitat of *A. myriacantha* at Coldspring near Grahamstown required much thought and planning as it was necessary to illustrate the dominant railway line and the small aloe beside it.

Some of the drawings

Aloe albida and *A. craibii* – habitat near Barberton, Mpumalanga, looking towards Swaziland, with pine plantations

Initially I had trouble finding this site. I made a slight mistake with the directions and arrived at what I thought was the site, did sketches, climbed a steep slope and reached the summit which looked over the really steep hilly landscape near Barberton. Eventually I found the correct site. I wanted to portray in the drawing the rock-strewn slope where the aloes were growing and the sea of alien plantations below it extending far into the distance. The composition was really based on a series of curves made by the hill in the foreground and complemented by the flowing, rolling forms of the plantations.

Aloe cooperi – Leolo Mountains, Sekhukhuneland, Limpopo Province

This was the first grass aloe landscape I illustrated. The site is dominated by Tama Kgoshi, the highest peak in the Leolo Mountains. I wanted to get a feeling of the steepness and ruggedness of these mountains, with grasses growing out between the boulders. There was a small patch of indigenous forest just out of view of the habitat depicted in the drawing. I wanted to include this, but after doing some preliminary drawings, I found that it did not fit into the layout.

Aloe linearifolia – outskirts of Margate, KwaZulu-Natal

This aloe grew in the vicinity of suburban dwellings surrounding Margate on the KwaZulu-Natal south coast. I wanted to include these buildings to show the modern-day environment of this aloe. To achieve the best composition, I had to look around extensively, but the problem was with a narrow road running near the site and no place really to pull over in a vehicle and get a good look. I parked at a nearby shop and walked into the coastal grassland where the plants of *Aloe linearifolia* were growing. The aspects I wanted to show in the illustration were the rolling slopes interspersed with giant *Strelitzia nicolai*, subtropical bush and the view into suburbia, with the sea in the distance.

Aloe micracantha – Bridgemeade, Port Elizabeth, Eastern Cape

The site where *Aloe micracantha* grows, borders suburban Port Elizabeth. The aloes grew in a very flat landscape. The plant was quite a nice, large specimen so I used it to dominate the foreground of the drawing, while the landscape extended off into the distance, with the long horizon providing the dominant feature. The aloes were quite far apart, so the scene here is basically as one would see the habitat. If you came across an aloe in one spot, you would probably not see another in your immediate field of vision.

Aloe micracantha – Greenhills, east of Grahamstown, Eastern Cape

This is a really beautiful location, with the aloes growing along the hill slope, long grasses, boulders and the prehistoric-looking *Oldenburgia grandis*. After climbing around on the slope, looking at the plants and getting a feel for the site, I ended up with a composition showing the ridge on the far side of the valley and hills in the distance. The foreground slope was dominated by the presence of the oldenburgias. In the immediate foreground I wanted to create a feeling as if the viewer were standing there. There is an aloe in the front of the drawing but it is quite hidden, showing how they occur and how they are not always that easy to see.

Aloe myriacantha – growing in the railway-line reserve near Coldspring, Grahamstown, Eastern Cape

In this drawing I really wanted to show the railway-line reserve, and had to manoeuvre around quite a bit until it all came together. To achieve this, I wanted the railway line to be shown quite well. As there was just a hint of it from a lower angle, it would only show up as a single dark line and the viewer would probably not know what it was, resulting in a drawing that did not convey the correct information. I ended up with the railway line being quite dominant and was initially worried that it would dominate the drawing. So, as I drew, I worked up different areas to make them stronger. The result is a balanced finished drawing. All the aspects of the site are there: the tiny aloes (looking large in the foreground), the rolling hills in the background, the storage dam in the distance, a few scattered aliens, the railway line, and the secondary road running parallel, below.

Aloe nubigena – Graskop Escarpment, Mpumalanga

This is an amazing site, looking down from the grassland escarpment on the highveld into the hazy distance of the lowveld. I wanted to depict the aloes hanging from the sides of the vertical cliffs, watered by the mists. I had to manoeuvre into quite an awkward position to get the angle I wanted, which was the steep rocky face, with the plants dangling from it, looking into the misty distance with The Sentinel towering out of the gorge.

Aloe verecunda – Carletonville area, western Gauteng

Near Carletonville this plant grows along a rocky ridge surrounded by flat plains, dominated by mining activity below. Mine shafts tower up above the surrounding plain, and mine dumps create artificial ridges and hills. Tall alien trees grow around the mines, also changing the landscape. All these features were taken into account to illustrate how this aloe fitted into this modified landscape of today. The ridge overlooking the plains is covered with interesting rocks of complex shapes, each one containing a mini-set of caves, convolutions and natural pools, all-in-all aesthetically pleasing. I wanted the illustration to include a foreground weighted down with these rocks, interspersed with grass, and an indication of an aloe somewhere in the foreground, but not necessarily in flower. In the middle ground, I wanted an indication of the ridge sloping off, meeting the plain below. In the distance I drew the mine shafts and mine dumps and the tall alien trees. The far distance fades off into the haze.

To achieve these perspectives, I drew a few quick sketches and then settled on the final composition. The complexity of the rocks and grasses posed a few challenges in the final drawing.

Aloe woolliana – growing in a seepage area on the Drakensberg Escarpment near Lydenburg, Mpumalanga

The aloes here grow just under the lip of the ridge. I walked around the habitat and sketched a couple of possible sites until I settled on the appropriate one, looking up the steep slope, with water seeping out onto the rock faces and the aloes nestled at the base of the grass clumps. I found a great spot to draw, leaning against a protea where I was very comfortable.

St Cyprian's Church – southwest of Grahamstown, Eastern Cape

The church is nestled in a lovely little valley. The main elements I wanted to portray were the building and the towering cypress trees, just next to it. I walked around the church and did some quick sketches to try and get what I was looking for. The area in the vicinity of the church is quite overgrown, so I had to clamber around in the bushes to get the angle I needed. I eventually found what I was looking for in fading light; it was quite eerie being alone, in the silence, with dusk falling and the two cypresses looming over the old stone church.

Taxonomy and the sociology of botanical knowledge

There have been two modern classifications of grass aloes in South Africa, that of Reynolds (1969: 110–155) and Glen & Hardy (2000: 4–22). The taxonomy used in this work is that of Reynolds. The main reasons are that the system used by Reynolds is not only a better reflection of the infrageneric taxonomy of the group, but also best suited to the elucidation of processes concerning the autecology of South Africa's grass aloes, which is the focus of this book. The classification used by Glen & Hardy takes its place in the knowledge production process as a more abstract model than that proposed by Reynolds. The recent work of these two authors is in keeping with current more abstract trends in the development of South African botanical knowledge. Such trends are to be expected in the history of the development of botanical knowledge in South Africa.

This chapter concerns a brief overview of the socially embedded process of knowledge production in South African botany known as the 'sociology of knowledge'. It is the author's conviction that this discipline is essential for understanding how classification systems work as they do. It is also useful in accounting for the regular revision of plant genera and changes in the names of species.

Families, genera and species are concepts defined by the purpose they serve in organising and classifying knowledge. These concepts in botany need not be demonstrated empirically and their historical use in taxonomy is probably closest to an art form rather than a science. Scientific methods are used to aid taxonomy based on the concepts of family genus and species. Chemotaxonomy and DNA sequencing are two examples of such techniques. None of these techniques, however, constitute plant taxonomy for as long as the present concepts of family, genus and species remain central to the process of hierarchical classification.

If scientific techniques are elevated to the level of taxonomy itself, then the concepts of family, genus and species will need to be re-theorised. Any changes in the status of these concepts can be expected to precipitate sharp debates in the botanical community, and the results of these debates can be expected to provide new directions for the development of botanical knowledge.

Publication of plant names is usually based on a description rather than the empirical proof of the existence of the entity. It is for this reason that the concepts of genus and species are likely always to be at variance amongst different botanists. The taxonomic choices that botanists make are influenced by the intellectual training they receive at tertiary institutions. This training is later broadened by working with leading exponents in particular kinds of taxonomic research.

In the early history of South African botany the emphasis was primarily on classifying the species diversity in different genera. Once names had been assigned to different entities in these genera, revisions became necessary. These have frequently been for the requirements of postgraduate degrees, supervised by an expert in the field of botanical knowledge concerned. The knowledge produced in the revisions has usually led to a systematic organisation of species names within a refined or redefined concept of families and genera. The tendency is for these newer models of classification to be produced at a higher level of abstraction than in the earlier phases of knowledge development.

One of the most significant aspects of plant taxonomy is that it used to be integrally related to its practical value for field application. In this sense, the sections of the genus Aloe that Reynolds proposed to classify grass aloes are typical. They reflect the closeness between the taxonomic system and the observable reality of grass aloes growing in the grasslands. The more abstract model used by Glen & Hardy collapses all South Africa's grass aloes into one section, which also includes *A. bowiea*. This species is, however, unlikely to be a grass aloe if one uses other classificatory criteria, as discussed later in this chapter.

Most taxonomic revisions reflect the consequences of the choices the taxonomist makes for organising knowledge. Anomalous data are quite often dealt with by reducing a number of names to synonymy under one species. This usually occurs where the concept of genus has the fewest criteria for its definition. In such instances the onus is placed on a later revision to redefine the concept of genus and reinstate some of the names formerly reduced to synonymy under one species.

The tendency towards constructing abstract models for plant classifications should place an onus on the taxonomic theorist to check the revision for internal consistency in the proposed model. The more the classificatory dimensions against which the model can be checked, the greater the internal consistency that is likely to follow. High levels of names reduced to synonymy under one species are sometimes, but not always, an indication of an impoverished concept of the infrageneric complexity of the group in question.

In the coming years, botanists are likely to need to reflect on and review the purposes of taxonomy. Matters that will merit consideration are whether taxonomy should be user-driven or become more abstract, designed to accommodate advances in theoretical thinking and debate.

Once a taxon has been defined and described, its name enters common language usage. The process of naming not only gives a plant its identity, but is also the label around which information about it is exchanged between people and in print. There is a social resistance to changes in the names of plants since each change necessitates a re-orientation in thought about the species and, in some cases, the genus in question. For this reason, amongst others, new names are often slowly accepted into regular language usage. If, on the other hand, names are understood as markers in the knowledge production process, there need be little resistance to them changing over time.

Some of the concepts introduced in the preceding paragraphs can now be discussed with reference to the grass aloe taxonomy proposed by Reynolds on the one hand and by Glen & Hardy on the other.

Reynolds's (1969: 104, 110–155) sections *Graminialoe* and

The habitat of *Aloe micracantha* in the vicinity of the farm Greenhills near Grahamstown in the Eastern Cape. The steep hill slope is dominated by *Oldenburgia grandis*. The aloes occur thinly in this habitat, usually in shallow soil amongst rocks where they experience least competition from other plants. The region has been heavily transformed by alien trees. A grove of invasive black wattles is visible on the hill in the far distance.

Leptoaloe of the genus *Aloe* are based on descriptive visible characteristics of the plants. They include all the grass aloes in South Africa. The names in these two sections include, in South Africa, all species that have deciduous leaves and grow in relationship with fire cycles. The strongest of these characters is their deciduous habit. *A. nubigena* and *A. thompsoniae* are subject to the full effects of fire cycles only where they grow in rocky mountain-top grassland. Plants growing on cliff faces are mostly, but not entirely, exempt from the regular effects of fire.

Glen & Hardy (2000: 4–22) restrict grass aloes to section *Leptoaloe*. They also include in this section *Aloe buettneri* and *A. bowiea*. Only the former is seasonally deciduous. In tropical Africa, Reynolds (1966: 9–47) includes some *Aloe* species in section *Leptoaloe* that are deciduous or semideciduous and that may or may not be subject to fire cycles.

If biological processes are used to regroup grass aloes, several species will need to be removed from section *Leptoaloe* and placed elsewhere. It could be more useful to create a new section of the genus *Aloe* in which to accommodate the many species with seasonally deciduous or semideciduous leaves and subject to fire cycles as an integral part of their biology.

It is accepted that grouping grass aloes together on the basis of biological processes may be viewed by some as artificial. On the other hand, it is also correct to argue that the natural classification systems for which taxonomists strive, are also constructs, since their authenticity is difficult to demonstrate empirically. Classifications of grass aloes based on biological processes are certainly valuable for preserving fire-based grassland ecosystems.

The discussion about the sections of the genus *Aloe* that accommodate grass aloes indicates some of the choices and problems taxonomists encounter at the conceptual levels of knowledge construction. It is legitimate to argue that, if specific biological processes can be used to group grass aloes into sections, equivalent processes need to be examined in other sections of the genus as well. This would lead to conceptual consistency in the knowledge construction process.

Grass aloe species names used by Glen & Hardy differ from those adopted by Reynolds in the following respects. *Aloe chortolirioides* is divided into two varieties, namely variety *chortolirioides* and variety *woolliana*. *A. dominella*, however, is retained as a species but the external visible characters of this plant are so close to those of *A. chortolirioides*, according to Glen & Hardy's description, that they hardly justify separation at species level. *A. boylei* and *A. hlangapies* are reduced to synonymy under *Aloe ecklonis* (Glen & Hardy 2000: 19–20), a decision that can be viewed as consequence of the concept of grass aloe presented at a level of abstraction excluding biological process. The concept of *A. minima* includes *A. parviflora* and *A. minima* var. *blyderivierensis*. Once again the level of abstraction excludes biological processes and significant observable differences in the field. *A. cooperi* is divided into subspecies cooperi and subspecies pulchra (Glen & Hardy 2000: 21–22). However, the same level of variation is found across the ecotypes of various *A. cooperi* populations. In addition, if *A. minima* var. *blyderivierensis* is not accepted, *A. cooperi* subsp. *pulchra* does not appear to justify retention in terms of the consistency of the model that Glen & Hardy propose.

The reasons for reducing described species to synonymy should be explicit and related to the internal consistency of the proposed model for classification. In terms of the process of knowledge production, lack of this consistency usually precipitates a later revision of the model.

Some problem areas in grass aloe taxonomy as it exists at present, are discussed below. Several of the populations of *Aloe vossii* on the Soutpansberg are morphologically very close to *A. verecunda* growing in western Gauteng near Carletonville, and to plants at Athalia near Piet Retief in Mpumalanga. Future research may reveal that *A. verecunda* is a very widespread species with several ecotypes, including *A. vossii*. *Aloe minima* var. *blyderivierensis* has several minor differences from *A. minima* var. *minima* in terms of its morphology and autecology. This variety has now become so isolated form typical *A. minima* populations to the south by plantations of alien trees that it may be useful to uphold its varietal status only for these reasons. There is a probability that populations of *A. chortolirioides* and *A. woolliana* may have intergraded with one another in former times before the advent of extensive afforestation between Barberton and Graskop, in Mpumalanga. Evidence for this proposition is to be found in two populations of aloes that have characters of both species. One near Carolina in Mpumalanga has the typical leaves of *A. woolliana*, but the flowers of *Aloe chortolirioides* var. *boastii*. The other population on the Steenkampsberg, also in Mpumalanga, is morphologically much the same as *A. woolliana* in terms of size and growth habit, but sometimes has flowers close to those of *A. chortolirioides*.

If a decision is made in a future revision to do away with the varieties of *Aloe chortolirioides* and place them in synonymy, the position of *A. dominella* would have to be carefully considered in terms of the model proposed. It is possible that *A. chortolirioides* and *A. woolliana* are extreme points in what was once a natural pattern of variation across the distribution area frequented by these two entities.

The knowledge construction principles that follow from the botanical concepts of genus and species imply that the process of naming plants will often change. Names, however often they may change, are an indispensable tool for all disciplines related to the study of plants. Not only do they serve as pivotal points in the construction of botanical knowledge, but they also allow knowledge-producing communities to assess the status of families, genera and species at given points in time.

The status of botanical knowledge-producing processes affects end users of classificatory systems in different ways. Disciplines concerned with systematics and taxonomy deal with the internal logic of classification systems. This is often in contrast to interest groups that use plant names. Their use of the names is driven by entirely different motivations such as compiling Red Data lists, drawing up plant appendices for CITES and drafting nature conservation legislation for the protection of flora. These user groups require certainty in plant nomenclature but for the reasons discussed, this requirement is unlikely to be met. Subdisciplines of botany such as autecology are differently placed with respect to the use of plant names. Workers in these fields can exercise choice as to which alternative revisions of plant genera best suit their purposes.

Workers in subdisciplines of botany will increasingly be placed in a position where they can choose from a range of revisions of different genera. The latest revision may not always be the most suitable for the purpose at hand. People who have an affinity for analysing the methods with which plant revisions are constructed and some formal or informal grounding in philosophical debate, will be placed at an advantage when choosing between competing taxonomic models.

Afforestation and the spread of alien plants in grassland

The grasslands of the Eastern Cape, KwaZulu-Natal, Mpumalanga and Limpopo Province have been heavily planted with alien timber trees, especially over the last century. Most plantations of alien trees are formally developed and managed by timber companies. In addition, large tracts of land have become infested with invasive Australian plants, chiefly different species of wattles (*Acacia*), often from the days of the leather-tanning industry, which used wattle bark for tanning leather.

Commercial timber plantations are in essence a managed type of crop farming. Invasion by aliens, often from timber plantations, is random and has, until the passage of recent stricter legislation, continued unchecked. The elimination of aliens from grassland is a huge undertaking which would have to be sustained over the coming decades if it is to have any long-term effect. The current elimination of aliens is frequently related to the rehabilitation of water catchment areas. Grassland species, including several of the grass aloes, are set to benefit from these initiatives.

Many plantations were established before there was an interest in preservation of grasslands and their constituent flora. Grass aloes have fared badly with afforestation programmes and many of the existing populations will require active management if they are to survive. Timber companies can be significantly involved in this process where interventions are merited and sustainable in the long term.

The process of afforestation and its effect on grassland ecosystems is a complex subject. The data discussed in this chapter have been collected over the last 20 years, mostly in the provinces of KwaZulu-Natal, Mpumalanga and Limpopo.

Some grass aloe species are either found only in areas of timber plantations or their distribution is nearly circumscribed by the plantations. Among these species are *Aloe albida, A. chortolirioides, A. integra, A. minima* var. *blyderivierensis, A. vossii* and *A. woolliana*. The newly described *A. craibii* is also currently known only from an area of alien pine plantations near Barberton in Mpumalanga. Grass aloe species for which an estimated 50% or more of the region in which they grow coincides with timber plantations, are: *A. boylei, A. hlangapies, A. kniphofioides, A. kraussii, A. minima, A. nubigena, A. saundersiae, A. soutpansbergensis* and *A. thompsoniae*.

Aloe nubigena, A. soutpansbergensis and *A. thompsoniae* are mainly cliff dwellers. In the case of *A. thompsoniae*, plantations have recently been cleared from the immediate periphery of the Wolkberg but some Australian plants still persist as invasive weeds. In the case of *A. nubigena* and *A. soutpansbergensis*, plantations of aliens above the cliffs where the aloes grow, draw off large amounts of ground moisture. This has led to some populations of these aloes inhabiting drier areas than in former times. Moisture for these cliff dwellers used to be provided by extensive seepage areas above the cliff summits, soft penetrating rain from dense mists and also heavy rainfall. Mist and rainfall are now the most reliable water supplies and many former seepage areas have been substantially degraded. Today these two grass aloes are confined mostly to sites that retain moisture the longest after heavy mist and rainfall in regions where there is extensive forestry activities on the mountain summits.

Aloe cooperi and *A. dominella* coincide to a minor extent, at present, with areas of afforestation. A more serious problem is invasion of the grassland with *Acacia mearnsii* (black wattle), an alien from southeastern Australia and Tasmania. *Aloe linearifolia* has the misfortune of growing in an area of

Above: The mountainous grassland habitat of *Aloe albida, A. chortolirioides* and *A. craibii* between Barberton and the Swaziland border in Mpumalanga shortly after a winter grass fire. The aloe habitat is confined to fragmented strips between vast tracts of eucalypts and pines. A patch of indigenous forest is visible in the middle distance.

KwaZulu-Natal that is largely under a monoculture of sugar cane and where this is not the case, much of its habitat consists of *Eucalyptus* plantations.

Aloe modesta also grows in regions of afforestation, but the biggest recorded population of this rare species is in high-altitude grassland which at present is not yet planted up with aliens. However, invasive *Acacia mearnsii* is beginning to become a problem and, if unchecked, could transform this important habitat.

Many of the significant habitats where *Aloe micracantha* and *A. myriacantha* used to grow in the Eastern Cape have been transformed by invasive *Acacia mearnsii* and *A. saligna* (Port Jackson willow), another aggressively invasive Australian plant. This process is continuing and remnants of former populations of these plants are restricted to rocky outcrops where the invaders cannot take root and develop into mature trees.

Aloe verecunda occurs mostly away from mountainous mist belt regions favoured for afforestation. Some of the habitat frequented by *A. fouriei* could have been planted up with alien timber trees, but for reasons concerned with South Africa's social history this has not taken place. The greatest proportion of *A. ecklonis* populations also occur outside timber-producing regions. *A. inconspicua* is not currently known from areas that support timber plantations.

Only three of the 26 grass aloe species are at present exempt from the negative impact of alien timber plantations and from the spread of aggressively invasive introduced plants into the veld. It is evident from the above discussion that the lives of most grass aloe populations of nearly all the species are lived out within districts of managed timber plantations and transformation of the original natural grassland by these practices. Various forms of intervention are required to preserve these plants in the medium term, but in the long term they are likely to become extinct. The management of grass aloe populations in timber-producing areas has to be directly related to the level of habitat alteration in which various grass aloe species are found. The discussion that follows is a more detailed consideration of these matters, drawn from some 20 years of field observations.

In the late 1970s and early 1980s much of the grassland in the Piet Retief District of Mpumalanga had already been destroyed by plantations of mainly eucalypts and pines. Some significant populations of *Aloe hlangapies* and *A. kniphofioides* were found in this area as well as a relatively pristine colony of *A. integra* near the small town of Amsterdam, close to the Swaziland border.

A number of these populations were found along streams in rocky grassland. These areas represented the low-lying parts of a former much more extensive habitat, now mostly covered by monocultures of aliens. In some regions, mostly with a large proportion of rocky ground, no afforestation had taken place. These were limited tracts of land that rarely adjoined unforested areas along streams. In a few cases of rocky hills with streams running below them, the full range of habitat occupied by *Aloe hlangapies* and *A. kniphofioides* was unintentionally preserved.

The range in size of unplanted habitats varies considerably, from small stretches of land only a few hundred meters wide to large spaces a square kilometre or more in extent. These populations provide ideal opportunities to assess the viability of various grass aloe populations under conditions of severe habitat alteration. The most viable colonies of plants in terms of the preservation of ecological processes are usually those in the most extensive habitat. This is so because many of the habitats characteristically occupied by the plants are still available across the general habitat. The plants are also sufficiently abundant in a large enough area to attract some of their usual range of pollinators.

One particularly large colony of *Aloe kniphofioides*, numbering several thousand plants, grew over a stretch of some 2 km of rocky grassland. This area, around a stream and below extensive plantations, varied in width

Above: Hilly coastal grassland on the outskirts of Margate, a town on the KwaZulu-Natal south coast, is home to *Aloe linearifolia*. Some alien plants such as pines are visible encroaching in the grassland. The bush clumps are dominated by *Strelitzia nicolai* and occur mostly along drainage lines, occasionally in the open veld. The Indian Ocean forms the horizon.

from about 100–700 m. During the 1990s much of this area was planted up. A once extensive and viable colony of *A. kniphofioides* was reduced to the few plants that occurred on rocky outcrops. In former times, pollination rates were excellent, based on the ability of the plants to attract a range of pollinators, particularly several highly active nectar-seeking species of sunbirds (genus *Nectarinia*). Currently a negligible number of plants are pollinated and there is little or no recruitment of new plants to the populations. In the medium to long term these remaining plants of *A. kniphofioides* can be expected to die out.

The policy of establishing plantations away from the grassland along streams certainly helps to effect the long-term survival of *Aloe kniphofioides*. In overall terms, this is a relatively small area and timber companies should be encouraged to preserve remaining areas of this habitat where *A. kniphofioides* is still to be found.

Aloe hlangapies grows less gregariously than *A. kniphofioides*, but its large capitate racemes of apricot-coloured flowers are very successful in attracting pollinators in the few areas around Piet Retief where the plants are still well represented. One rocky area northwest of Piet Retief still supports a large colony of *A. hlangapies*. The undulating land around the rocky hills is completely covered with plantations. The *A. hlangapies* colony extends over about 2 km^2 of rocky grassland. The plants are conspicuous in October and November when in flower amongst the short grass. They still manage to attract good numbers of black sunbirds (*Nectarinia amethystina*) and malachite sunbirds (*Nectarinia famosa*). Apart from the aloes, this patch of grassland is rich in other flowering plants attracting other general pollen- and nectar-gathering insects such as honeybees. In some years, smaller local hills close by also attract pollinators since the general area consists of a patchwork of natural and artificial habitat. This is in marked contrast to many places near Piet Retief where there are occasional small patches of veld within a vast monoculture of eucalypts and pines.

Close to the hamlet of Amsterdam is a remarkable site where several species of grass aloes are growing in close proximity to one another. It is a valley about 4 km long, with hilly and rocky grassland on either side of a river. This locality is significant in that it represents the full range of habitat diversity available to the various species. The site currently provides an important opportunity for research purposes since invasion of the hills by *Acacia mearnsii* is taking place on one side of the valley and the initial stages of commercial timber afforestation on the other side. The effects of the destruction and alteration of the habitat can be studied over time, with reference to specific grass aloe populations.

Aloe kniphofioides occurs on the flatter ground towards the valley bottom and also on the sunny west- and southwest-facing slopes. The plants grow in rocky areas of short grass on the hillsides as well as in patches of denser grass with deeper soil. *A. integra* is restricted to the west- and southwest-facing slopes. *A. boylei* grows on steep banks along seasonal streams and the perennial river. This plant is common only on the south- and southeast-facing slopes and largely in the vicinity of moisture-retentive gullies and seepage areas. In all, these grass aloes occupy less than a quarter of the total habitat available. The significant point is that they occur across all possible habitats except for rock faces and barren rocky outcrops inhabited by succulents such as *Euphorbia pulvinata*.

In addition to the destruction and alteration of the habitat at this locality at present, the hills are heavily grazed by sheep, cattle and goats. This phenomenon is typical in areas heavily afforested with aliens, where rangeland for domestic stock becomes progressively more limited. Many of the larger grass aloe species in these afforested areas are sufficiently hardy to withstand periodic trampling by livestock. The problems really arise when the aloe inflorescences are snapped off and trampled, leading to a reduced seed bank in the environment.

The Barberton area of Mpumalanga once was one of the richest habitats for grass aloes in South Africa, but the region between Barberton and the Swaziland border has now been destroyed by extensive plantations of eucalypts and pines. The habitats that remain are mostly fragmented patches of mountain grassland of varying size. With the addition of further plantations and the spread of invasives, some of the remaining habitat has shrunk over the years. There are two significant nature reserves in the region in which some of the species can be conserved, but most plants are found outside these reserves on land owned by timber companies.

Aloe chortolirioides used to be widespread in the mountains between Barberton and the Swaziland border. A few viable populations remain as they are situated on the fringes of the main forestry region. These populations are subject to regular burning cycles unrelated to managed firebreaks associated with timber production. The plants are also widely scattered across large areas of ideal habitat, attracting the full range of pollinators when they flower *en masse* after winter grass fires.

Smaller populations within the timber-producing area are subject to infrequent fires and many years may pass in succession with no fires at all. The habitat lies within a fire prevention area protecting timber plantations. These plants are in effect removed from the fire cycle, which is essential for maintaining their life processes. On occasions when fires do occur, flowering performance is often poor and seed set limited owing to the limited number of flowers and paucity of pollinators. Infrequent fires favour the flowering of plants in open, exposed, sunny places whereas plants on shady, infrequently burnt slopes often produce a small number of flowers. Plants on shady slopes fare badly when burnt irregularly. The older stems are prone to rot and fungal attack when surrounded by moribund grass for successive years, particularly during summers with prolonged mists. Fire maintains healthy, compact plants and its exclusion represents a major problem for the long-term survival of *Aloe chortolirioides*.

The stronghold of *Aloe albida* used to be rocky outcrops and associated short grassland on the mountains between Barberton and the Swaziland border. The short mountainous grassland frequented by the plants is now largely forested with timber. Populations that remain, are now found mostly on rocky outcrops and mountain summits, too rocky for planting timber. The absence of regular winter grass fires has further shrunk the habitat of these plants. An additional problem is that some of the few remaining habitats are currently being transformed by invasions of *Acacia mearnsii*. The species occurs in neighbouring Swaziland, but the problems associated with survival under conditions of afforestation are repeated in South Africa.

Aloe craibii is currently known from a few small colonies in the Barberton area close to the border with Swaziland. Part of the population at the type locality has been afforested with pines during the last decade and the mountainside at another locality nearby is in the initial stages of transformation by black wattles. Pollination of the plants at the type locality is erratic. The area is surrounded by aliens, creating unsuitable habitat for pollinating agents such as sunbirds. The usual problems associated with the exclusion of regular fires are also repeated for this species.

The hilly region around The Brook, near the western boundary of Swaziland with Mpumalanga, is heavily afforested with pines and eucalypts. Some areas, however, are only partly under timber. The parts not under plantations are extensive, usually rocky hillsides on either side of broad valleys. This region is one of the strongholds of *Aloe minima*, a species partial to hilly, rocky grassland. The broad corridors of habitat between timber plantations contain large viable populations of *A. minima*. Winter grass fires sweep up the valleys, particularly where they border a densely settled region inhabited by rural pastoralists who regularly burn the veld. The plants grow well and most flower for a two-year period after a grass fire, with flowering performance gradually decreasing until the next fire. The seed set is good during years conducive to favourable flowering.

Afforestation has not seriously affected the life cycle of *Aloe minima* in the vicinity of The Brook. This is entirely due to the manner in which afforestation has been excluded from much of the habitat frequented by *A. minima*. In addition, the pollination biology of the plants has not been significantly interrupted as it has in most areas of afforestation.

Most grass aloe species are found in areas of managed timber plantation and some grow in regions where the grassland has been invaded by introduced Australian acacias, particularly *Acacia mearnsii* and *A. dealbata*. Legislation governing the spread of these plants as well as initiatives for their control and eradication has recently been passed in South Africa. These measures, if properly applied, ought to promote the preservation of the habitat in which grass aloes are found. In some cases, however, the habitat of rare grass aloes can be transformed within as little as a decade. A case in point is the habitat of *Aloe modesta* near Wakkerstroom in Mpumalanga. *A. micracantha* and *A. myriacantha* in the Grahamstown and Port Elizabeth Districts of the Eastern Cape are currently the two species whose future existence is jeopardised by transformation of the grasslands by invasive alien plants.

In timber-producing regions and in areas of grassland extensively invaded by aliens, there is sometimes extensive hybridisation where significant numbers of two species occur together and flower simultaneously. This affects mainly larger grass aloe species pollinated chiefly by sunbirds. These bigger populations are usually the most viable in areas of grassland destroyed and altered by aliens. Sunbirds visit huge numbers of aloe flowers in the open veld. They generally move over large populations of flowering aloes without repeatedly visiting the same flowers during one feeding event. In afforested areas, sunbirds frequently visit the same flowers during one feeding event, lasting a few hours. This feeding method greatly enhances the chances of pollen transfer between species.

Two fertile hybrids have been observed as a result of grass aloe species flowering simultaneously and attracting the same pollinators: *Aloe boylei x A. kraussii* and *Aloe ecklonis x A. hlangapies*. The hybrids both occur in regions heavily forested with aliens near Greytown in KwaZulu-Natal and close to Piet Retief in Mpumalanga. In areas around Piet Retief, close to the Swaziland border, that have escaped afforestation, *A. hlangapies* still occurs with its diagnostic distichous leaves and densely capitate racemes of apricot-coloured flowers.

It is fortunate that most grass aloes that grow together in the same grassland usually have different flowering seasons. The situation is more complex where species grow together and their flowering times overlap. In such instances, hybridisation brought about by artificial conditions has to be taken into account when grass aloe populations are identified for long-term preservation.

The preservation of most of South Africa's grass aloes in areas with commercial timber and grassland invaded by aliens depends on two related factors: the identification of viable populations of the different species and management of many of the populations that fall within timber-growing areas. Particular attention ought to be given to the preservation of the very few grass aloe populations where most of the complex biological processes that regulate their life cycles are still intact. Initiatives in this regard will require the co-operation and commitment of timber companies as well as a range of landowners and land users to ensure even a limited amount of success.

The changing landscape: mining, urban development and informal settlements

Above: The habitat of *Aloe verecunda* near Carletonville. The hills around Westonaria near Carletonville are still a stronghold of the species despite extensive development in the area. Rocky outcrops on the summits of these hills are often densely utilised by the aloes. Mine dumps and a mining headgear are visible in the distance.

The South African landscape has been significantly changed and transformed since the arrival of settlers. Much of the northern hemisphere from which most of the settlers originated, consists of man-made environments dominated by the requirements of urban living, agriculture and industry. Technology and science have for a long time been the dominant world ideology and this ethos was transported to South Africa with European concepts of the ownership of productive resources. The indigenous people living in South Africa's grasslands had little serious impact on the grassland ecologies that took millions of years to establish.

In South Africa, as elsewhere in the Western world, progress is still largely equated with advances in science and technology. In this hierarchy, grasslands have been used and managed to suit agricultural and industrial interests. This chapter and those that follow examine how human interests have shaped the environment in which grass aloes live and reproduce.

Open-cast mining and deep-level gold mining are practised in many habitats occupied by grass aloes. There are numerous deep-level gold mines in western Gauteng and the adjacent North-West Province, and widespread coal mining over much of the Highveld grasslands in western Mpumalanga. Gold mining in western Gauteng and eastern North-West has been beneficial for the preservation of some of the most significant concentrations of *Aloe verecunda* in South Africa. Coal mining has, in contrast, destroyed some of the habitats of *A. ecklonis*. Open-cast mining, such as the coal mining in Mpumalanga, destroys much of the land surface locally, through extraction of the coal and dumping of overburden. *A. ecklonis* is one of the more common grass aloes and the destruction of some populations in Mpumalanga is of little overall significance. It is important to note, though, that open-cast mining is much more destructive to the grassland environment than deep-level gold mining.

The gold mines in western Gauteng have laid claim to a particular type of land ownership. There is little grazing by domestic stock, insignificant development of crop farming and few, if any, artificially maintained barriers to winter grass fires. The stony ridges on which *Aloe verecunda* still occurs in healthy populations, are largely in pristine condition.

Before the advent of gold mining, much of the hilly area around Carletonville was used for marginal stock farming. Some areas were severely over-grazed and evidence of this is still to be found on some land owned by gold mining companies. These signs, prevalent almost a century later, include veld overgrown with *Seriphium plumosum*, a common greyish shrublet, and proliferation of common species of *Ledebouria* and *Albuca*. These naturally occurring bulbs become very common in heavily over-grazed areas since they experience minimum competition from grasses.

Gold mining unintentionally functions to preserve many grassland species in western Gauteng, including *Aloe verecunda*. The waste rock produced by shaft sinking and stope development as well as the by-products of ore extraction are dumped in designated areas. These are usually open veld well away from the stony ridges where most of the diversity of Highveld flora is to be found. Urbanisation connected to the gold mines takes place in limited areas and sometimes small herds of domestic stock are kept, usually a few horses and cows. These animals prefer to graze in flat, open grassland rather than on rocky hillsides.

Many ecological processes have been preserved by deep-level gold mining and forms of development inimical to natural ecosystems have been kept out. For example, indigenous antelope are still to be found in western Gauteng and these animals graze the grassy hillsides. In the dry winter months fires sweep across the habitat as in times gone by. The full range of grassland pollinators are usually present. The future of grassland ecological processes in areas where gold-bearing ore is exhausted merits attention from people concerned with the preservation of the grasslands when the gold mines are closed.

Urban and suburban development as well as the establishment of informal settlements on the periphery of towns, is currently transforming some significant grasslands and the ecological processes within them. Informal settlements are unplanned to a greater or lesser extent. Many of the inhabitants are unemployed and derive some of their food from livestock kept in areas adjacent to the settlements. This is in contrast to suburban development which is planned, and sometimes avoids rocky hills and ridges where several grass aloe species are to be found. Some instances of both types of development are discussed below, with examples of how various grass aloe species have responded to changes in the environment over the last two decades.

The south coast of KwaZulu-Natal used to consist of extensive tracts of coastal grassland, providing suitable habitat for *Aloe linearifolia*. Most of this has been destroyed to make way for sugar cane fields and plantations of alien trees. Intact tracts of grassland are largely found around some of the south coast towns such as Margate. The gently undulating hills around the town have until recently been home to some of the largest remaining viable populations of *A. linearifolia*. New suburban development is gradually destroying the flatter areas most suitable for this type of development. The plants are now found mostly on the rocky hills representing as they do only about half the natural habitat favoured by this species. Recruitment of seedlings to the populations is also slowest in the rockiest areas, making these populations particularly vulnerable to habitat disturbance.

It can be expected that as time progresses, the Margate populations will become increasingly artificial, occurring mostly where they are inadvertently protected by specific forms of land use. In this respect, the small fenced-in population around the Margate airport and the one near the water supply reservoir may eventually be all that remain.

Where urban planning accommodates reserves that include natural grasslands, grass aloe species fare well. Some of the best preserved colonies of the rarer species are those of *Aloe linearifolia* in the Kranskloof Nature Reserve. This reserve lies in the suburbs of Durban in KwaZulu-Natal and is an important vehicle for preserving a range of rare cliff-dwelling and grassland plants. The conditions under which *A. linearifolia* grows in the Kranskloof Nature Reserve exclude most of the human-induced changes that have occurred in the grasslands of KwaZulu-Natal. There is an absence of grazing by domestic stock and burning occurs in the dry winter. The conditions for pollination of the plants are favourable and there is a healthy recruitment of young plants to the population. The full range of grassland occupied by *A. linearifolia* is preserved, albeit at a reduced scale.

Not very far, as the crow flies, from the Kranskloof Nature Reserve is the Botha's Hill area. This area is an important suburban refuge for *Aloe parviflora*. Suburban development has avoided the areas with the rockiest grassland which represent the optimum habitat for this dwarf grass aloe. There is no grazing by domestic stock except for the densely settled region known as The Valley of a Thousand Hills. The few colonies here co-exist with heavy year-round grazing. Most of the plants are restricted to rocky niches where they escape trampling. This habitat represents a small fraction of that occupied by the plants in the well-preserved rocky grassland around Botha's Hill where the plants are relatively common and seed set is good year after year under conditions of average or above average rainfall. Individual habitats are often about a hectare or less in extent. This limited area is quite enough to preserve the ecological processes governing the life cycle of this diminutive grass aloe.

The general area is seriously invaded by stands of alien vegetation, which often goes hand in hand with urbanisation and habitat disturbance. The rocky grassland has been able to resist invasion by aliens as the soil is too shallow for many of them to take hold. In addition the grassland is healthy and not over-burnt.

The demise of the previous political dispensation saw a huge influx of people to South Africa's towns, looking for employment opportunities. Many of these informal settlements are built on areas of once well-preserved grassland. *Aloe inconspicua* used to be found in good numbers on the outskirts of suburban Estcourt in KwaZulu-Natal. The plants occupied the full range of habitat available. They occurred in sparse grassland around acacia trees and in the low-lying areas and also on stony hillsides where they grew amongst short grass. The low-lying areas are now fully developed with dwellings and the hillsides are heavily grazed by cattle, sheep and goats. Firewood is removed from the acacia thickets on the hilltops. A greater proportion of the habitat is in the initial stages of erosion and also invaded with weeds where there has been extensive disturbance of the surface soil.

Aloe inconspicua is still to be found on the rocky hillsides. It remains to be seen whether or not the plants will adapt to the changed conditions. In former times the area surrounding Estcourt preserved representative populations of *A. inconspicua* as well as the ecological process under which they grew. The future of the species is now in the hands of the farming community and their requirements for land use. In some places cattle ranching is giving way to game ranches in an effort to combat huge losses from stock theft. Game farms may well be a good unintentional method for preservation of the grassland and rare grass aloes such as *A. inconspicua*.

Leandra is a small town on the Highveld in western Mpumalanga. During the last decade there has been a huge influx of informal settlers to the area around Leandra and the adjacent towns. *Aloe ecklonis*, a common hardy grass aloe, used to be plentiful around Leandra. The plants were found in railway-line and road reserves and in undeveloped grassland on the periphery of the town. The fenced reserves kept out grazing domestic stock. *A. ecklonis* produces enormous quantities of seed and as a result large populations of this species established themselves in these reserves. In recent times wire fencing and fence posts have been removed by local inhabitants and the veld is once again heavily grazed. Most of the young plants have been trampled out by grazing stock. The flowers and developing flower buds on the mature plants are often snapped off and trampled by grazing animals. It can be expected that even a hardy grass aloe such as *A. ecklonis* will disappear from the grassland where it was once common.

Some of the larger grass aloe species grow comfortably alongside rudimentary pastoralism, but the smaller species usually do not fare so well. The responses of specific plants to pastoralism are discussed in the chapters concerning the various species. However, in the case of dwarf species there is a limit to the amount of grazing that can co-exist with healthy populations. *Aloe saundersiae* is a case in point. This species used to be common in the mountainous grassland around Nkandla in KwaZulu-Natal. The plants have declined markedly over the last two decades. They used to be fairly frequent on the rocky outcrops with sparse grass cover and on very steep slopes covered with dense, short grass. The aloes have all but disappeared from this habitat owing to heavy grazing by goats. These animals have trampled the soil which has become compacted and eroded. The dwarf aloes are now found only where they are protected by rocks and occur in isolated pockets, often far away from one another. The changes in the *A. saundersiae* habitat have been brought about by an expansion of the numbers of people living in the villages and stocking the veld with much greater numbers of animals. *A. saundersiae* is now mainly a rock outcrop plant in the Nkandla area.

The processes discussed in this chapter indicate how some forms of land usage are conducive to the preservation of grass aloes. It would be useful if modern urban planning could accommodate some formal grassland reserves and also give recognition to areas that are already conserved as a by-product of certain kinds of suburban planning.

Fire and burning in the habitat

The autecology of a grass aloe is inseparably linked to grass fires. The plants have not only evolved with fire but their very existence also depends on it. This chapter concerns a detailed discussion of the annual growth cycle of grass aloes and the timing of grass fires in relation to their growth habits. Widespread fires in pristine grassland usually lead to mass flowering the following growing season and, consequently, the successful attraction of the full variety of the plants' pollinating agents.

Much of eastern South Africa, which is now forested with alien timber-producing trees, forms part of an artificially managed grassland burning regime. These circumstances are largely different from fires that either are lit deliberately to stimulate early autumn or spring grazing or occur naturally in large tracts of open grassland. Many parts of eastern South Africa are lightening-prone, particularly elevated mountains and the high-altitude inland plateau.

Humans have always been part of the African ecosystem, and so has fire. Outside the Fynbos Biome, humans significantly increased fire only in those areas where they started practising summer burns, as in parts of Transkei and KwaZulu-Natal. This practice goes hand in hand with increased human population densities. In other areas, the opposite is probably true: in modern times humans have suppressed fire in agricultural and forestry areas. One only has to look at satellite images of Africa in winter to see how dramatically fires diminish south of the Limpopo River, especially in areas in private ownership. This is the result of the firebreaks often associated with fenced-in areas. In olden days, veld fires would have burnt over hundreds of square kilometres during the dry season, probably on an annual basis (although mosaics would have remained unburnt because of wind direction, rain, rockiness and other features of the local topography and vegetation). The suppression of fire in some timber-producing areas is a case in point. On the other hand, humans have dramatically increased the fire frequency in the Fynbos Biome, but that is of no relevance to the grass aloe story.

Present populations of grass aloes are subject to all the causality that underlies grass fires, whether natural or under artificial control. Some species have responded opportunistically to changes in the fire ecology and their populations have expanded. They have declined, however, when other unrelated alterations, such as changes in human settlement patterns, take place in their habitat.

It is necessary to reconstruct what probably were ideal burning regimes for grass aloes at the time the grasslands were sparsely settled by indigenous pastoralists. The reconstruction, which is likely to be only an approximation, is based on observations in the only few remaining areas of uninterrupted sparsely settled areas of grassland still available for study.

In former times grass fires were caused by lightning strikes and humans and had the greatest incidence of probability in the late winter and early spring. The grass is tinder dry at this time, windy conditions are frequent and huge areas of grassland could be set ablaze. The only natural barriers were rivers and mountain ranges with extensive cliff faces. Grass fires are also subtly directed by changes in the direction of the wind and by areas that have remained moist during a mild winter and where the grassland is only partially flammable.

Above: A veld fire near Dullstroom in Mpumalanga, with burnt and unburnt areas of grassland in the background and a line of flames burning the dry winter grass in the foreground. Some areas have escaped burning owing to subtle changes in the direction of the wind during the course of the fire.

Massive grass fires over large areas of pristine grassland followed by good early summer rains stimulate extensive flowering in the larger grass aloe species. This results in ideal conditions for the pollinators, from small insects to larger nectar-seeking sunbirds. These conditions were frequent before the advent of agriculture and alien timber plantations. Agriculture and the timber industry have suppressed the frequency and duration of fires. If the burning of the grasslands is followed by drought, few grass aloes flower, many succumb and limited numbers of pollinators are attracted to pollinate the plants. If widespread rains fall in the early summer without fires the previous winter, poor flowering conditions are prevalent but for different reasons. The synchronisation of fire, rainfall, high levels of flowering and the attraction of pollinators is indeed an intricate and delicate balance.

One of the most remarkable adaptations of grass aloes is that their dead winter foliage is always as ready to burn as is the condition of dryness of the grass in which they occur. The best time for making field observations is in June, around the time of the southern winter equinox. By this time, drops in temperature and successive frosts followed by warm, dry days have prepared the grassland for fire. In this respect observations were made on the dry leaves of *Aloe boylei* and *A. integra* near the small town of Amsterdam in Mpumalanga on 12 June 2004.

There had been good late summer and early autumn rains in the period late February to early April. The shaded south-, southeast- and southwest-facing slopes where *A. boylei* grows were still moist and the grassland had not died back fully. Some grassland plants such as *Geranium wakkerstroomianum* were still in full leaf. The *A. boylei* leaves had died back to about three quarters of their length. The leaf bases were still succulent and green.

The northern and northwestern slopes on the opposite side of the valley are exposed to direct sunlight for most of the day. The *Aloe integra* leaves had died right back to the main stems of the plants and the leaves were tinder dry, like the surrounding grassland. This habitat was ready to burn from shortly before the winter equinox until the first rains in October when the grassland resprouts after its winter dormancy. The grassland had not burnt for some years and should a fire occur, the *A. integra* plants had accumulated a good deal of leaf material for a hot burn.

The ideal timing of fires for the larger grass aloe species is late July and early August, just before the beginning of the growth cycle. Growth of grass aloe leaves begins, albeit slowly, with the increasing daylight hours in late July and August. Early grass fires leave the plants exposed to the direct rays of the sun for several months and also to the attention of grazing antelope and domestic stock. There is always some greenery at the leaf bases of larger grass aloe species. This is sufficient to attract grazing animals in an otherwise burnt landscape offering no alternative grazing. In years with late winter grass fires, the new grass sprouts at the same time as the grass aloes, in late August and September. The grass, even though available only in small quantities so early in the season, is usually grazed in preference to the aloes.

The early-flowering, larger grass aloe species initiate bud formation from the second half of September, often in the absence of spring rains. As the temperatures rise in October, the flower stems develop. Good early summer rains result in spectacular masses of flowers in grass aloe populations that have been subject to burning the previous winter.

Most bigger grass aloes flower after a fire, with their leaves and flower stems raised above the level of the fresh green grass. In this condition they are fully exposed to direct sunlight, required for the proper development their flowers. They are also at their most visible and conspicuous to pollinators.

The smaller grass aloe species, with the exception of *Aloe inconspicua* and *A. kniphofioides*, develop their leaves simultaneously with the main summer rains, from late November to late December. In most of the smaller plants the complete leaf development has been attained by the time of the southern summer equinox. The larger grass aloe species require a very open, moist, sunny habitat to grow and flower well. In contrast, the smaller species need to grow and flower amongst light grass cover. Conditions like this are usually at their best in the late summer after a winter grass fire. In very rocky habitats grass cover thickens over several years and is not ready to burn every season. This is ideally suited to the requirements of the smallest grass aloe species since light accumulations of decomposing grass shelter their roots from the desiccating rays of the sun at the peak of the flowering season in February.

The two small species of grass aloes, *Aloe kniphofioides* and *A. inconspicua*, have peak flowering periods in October in the case of *A. kniphofioides* and November in the case of *A. inconspicua*. Both have underground bulbs. *A. kniphofioides* requires burning to stimulate mass flowering since the plants occur in short but dense grassland. Flowering is greatly reduced in the inter-fire years, but abundant once more in the early summer preceded by winter grass fires. *A. inconspicua* rarely grows in very dense grassland. It flowers well without fire in sparse grassland but does require fire to grow and flower in habitat where the grassland is dense, albeit short.

A few grass aloe species rarely flower except after the grassland is burnt. The most notable is *Aloe chortolirioides*, which may not flower at all during years when the grassland remains unburnt.

One large grass aloe, *Aloe craibii* endemic to mountainous grassland in the Barberton area of Mpumalanga, flowers in early autumn, late March to the beginning of April. In the case of this species winter fires clear the habitat of dead grass, thereby improving the growing conditions for the plants in the summer months. This grass aloe is one of the few large species that frequently flowers during prolonged periods of misty weather.

It is evident that fire plays several varied roles in the lives of grass aloes. For some, fire is a direct stimulant for flowering. For others, perhaps most of the South African species, fire regulates the conditions under which they grow and flower.

Natural fires and those caused by humans represent the total expression of the types of fire possible in the grasslands. Natural fires and those caused randomly by people are erratic. Each grass fire in one locality differs from the next over successive years. Burns may be superficial or hot, and they may occur in successive years during dry winter seasons or only infrequently. In addition, some areas of the habitat may get burnt but others not owing to changes in the direction of the wind during the course of a fire. This random nature of most fires permits grass aloe populations to remain stable, expand or contract. In the case of many species an interval of two to three years between fires permits young seedlings to mature into their fire-resistant state. Their young trunks are sufficiently hardened off to resist destruction in all but the hottest burns in the grassland.

Successive annual fires over a long period, apart from changing the composition of the grassland, and promoting the growth of annual weeds, may cause a decline in the recruitment of new plants to grass aloe populations. The paradox inherent in these circumstances is that, although the plants may flower and seed well in successive years owing to annual winter grass fires, few seedlings escape destruction.

The more that fires are managed by human intervention, the less the possibilities for grass aloe populations to express their genes through a natural seeding and seedling recruitment process. This modern change has now become entrenched in the greater part of grassland habitat extensively utilised for timber production. Fire breaks have to be burnt early each winter to protect timber plantations. Smaller areas of grassland between plantations are often completely burnt, whereas in large tracts of land between plantations only strips of grassland adjacent to the timber are burnt. Both kinds of fires have negative effects on the long-term survival of grass aloes. Grassland that is burnt every winter generally supports large mature plants since opportunities for recruitment of seedlings to the population are greatly impaired. Grassland that escapes burning between fire breaks for successive years, is not conducive to the flowering of grass aloes or the establishment of seedlings. Seed that may be available for dispersal in the environment is unlikely to germinate under a canopy of dead grass without sufficient light and free air circulation.

Carefully managed burning programmes instituted by timber companies in the grassland between plantations could alleviate the problems that have been discussed. It is also true that the more an ecosystem is managed by humans, the more such interventions must be sustained over time. In the case of these grasslands, the problems in the habitats between timber plantations are different from those in other grassland habitat. The grassland is sometimes used for intensive grazing, it frequently becomes infested with annual and perennial weeds and it may also be destroyed by further afforestation. Efforts to preserve the plants are probably most productive in ecosystems where as few as possible of the biological processes have been disturbed by afforestation.

Fires and burning in timber-producing areas provide significant research opportunities for students of disturbed processes in grassland ecosystems. This knowledge could provide significant insights required for the preservation of general grassland ecosystems where grass aloes and other plants reliant on burning cycles are still to be found.

Two other kinds of fires occur in the grassland habitat—summer fires deliberately lit to stimulate autumn grazing and mainly annual winter fires around the informal settlements on the outskirts of many towns.

Autumn fires have significant negative effects on populations of large and smaller grass aloe species. They are usually lit in February and early March to stimulate autumn grazing. This is the peak flowering period for the smaller grass aloes and the species most at risk are those in open, short grassland. Buds and flowers often get burnt, but these fires are only possible if the grassland is dry. February is usually the wettest month in many grassland areas and flowering is usually at its best during years of abundant late summer rainfall. In the case of the larger species, the peak seeding season is usually from late January to early March. Autumn grass fires precipitate the premature dormancy of the plants and also destroy much of the season's seeds. Autumn fires are not lit every season, even if conditions are favourable for burning. For example, a large colony of *Aloe modesta* in the Dirkiesdorp area of Mpumalanga was burnt only once during the decade while the research for this book was in progress. Grazing requirements vary from one year to the next and in this sense the autumn fires are much less disruptive of ecological processes than managed fires in the timber-producing regions.

Fires around informal settlements potentially occur whenever the grassland is dry, and are consequently most frequent from May to early October. Informal settlements have been established on and around the colonies of several grass aloe species where they are in flattish areas suitable for the construction of dwellings. Livestock is grazed on the surrounding hills. The increasing frequency of fires has usually improved the flowering performance of grass aloe species that grow nearby. Increased grazing pressures associated with informal settlements keep the grassland short and minimise the effects of frequent fires. The impact of these settlements and their associated fires on grassland flora is a new phenomenon associated with urbanisation patterns in post-apartheid South Africa. Some flora can be expected to respond positively to increased fires, whereas the reverse may be true in other cases. At present there is insufficient information to assess the long-term effect that fires around informal settlements may have on grass aloe population dynamics.

Fire plays a critical role in regulating the conditions for the germination of grass aloe seeds. It keeps the environment free of moribund grasses and destroys excessive amounts of seed produced by some of the larger species such as *Aloe ecklonis*. Fire is associated with all aspects of the annual growth and dormancy cycle of grass aloes. The specific ways in which it regulates the growth cycle of individual species is discussed in the chapters that follow.

Habitats and growth cycles

The account that follows is a broad overview of grass aloe habitats and growth cycles. The themes discussed here are described in detail in the main treatments of the various species.

Most grass aloe species grow in regions of hilly or mountainous grassland. These areas afford the greatest range of habitat types that can be utilised by the different species. Rainfall is mostly in summer, from October to early April. In addition, most grass aloe habitats experience heavy mists in summer. These mists provide soft, penetrating rain for species that grow on cliffs or rocky mountaintops in escarpment areas. Grass aloes avoid the drier western grasslands of the South African interior.

The most common and widespread species make use of the greatest number of niches in a given habitat. Some of them, usually the more robust hardy species such as *Aloe ecklonis*, are also able to exploit temporary or permanent changes to particular grassland habitats.

In many plant communities in South Africa species are specifically adapted to sunny northern or northwestern aspects of a hilly or mountainous habitat. Different plant communities are found on shady moisture-retentive southern, southeastern or southwest-facing slopes. This habitat differentiation is not significant for grass aloe species that grow in regions that experience prolonged misty weather in summer. *Aloe albida* may, for example, occupy any suitable habitat in the mountaintop mist belt area near Barberton, as may *A. chortolirioides*, found in the same area.

Hilly areas with little or no summer mist have sharp divisions in habitat that is occupied by grass aloes. This is at its most evident in western Gauteng where there are large populations of *Aloe verecunda* on south-, southeast- and southwest-facing hill slopes. These plants are usually rare on the sunny, arid northern slopes. At Athalia near Piet Retief, *A. verecunda* is found mostly on shady slopes but also on the upper sunny northern crests of rocky hills. This area is in the mist belt and experiences high rainfall. These conditions afford this grass aloe the chance to occupy a wider range of territory in the overall habitat.

Grass aloes grow mostly in parts of a habitat with the shortest grass, except for the largest species such as *Aloe cooperi* and *A. ecklonis*. The bigger species have extensive root systems and their robust habit allows them to compete with taller grasses. They are also able to grow well on rocky hillsides where their habit is more stunted owing to the shallow soil thinly distributed around rocks.

A few grass aloes grow only or mainly in shady or semi-shaded habitats irrespective of their occurrence in a mountainous mist belt. In this respect, some colonies of *Aloe nubigena* are in complete shade throughout the winter in their cliff face habitats near Graskop in Mpumalanga. In summer they experience only a brief hour or two of direct sunlight in the early morning. In contrast, *A. integra* is mostly an inhabitant of dry, sunny, rocky hillsides.

Most grass aloe habitats have been severely altered and degraded by human activities. These alterations have been brought about by heavy utilisation of grasslands for afforestation with alien timber trees, agronomy and stock farming. The range of habitat available for particular species has shrunk in many cases. Few grassland areas remain where individual species are able to occupy the full range

Above: The boulder-strewn grassland habitat of *Aloe hlangapies* near the Houdkop border post between Mpumalanga in South Africa and Swaziland in the middle of spring in 2004 when the veld was ablaze with a wealth of flowers. The aloes are largely absent from bush-clad hills such as those in the background, preferring the open habitat depicted in the foreground. Areas such as this still contain a wide diversity of flowering plants in many genera.

of habitat available to them in former times. These circumstances restrict the opportunities available to grass aloes as a whole to adapt to environmental changes such as global warming.

Grass aloe growth cycles vary with the size of the species. All the smaller South African grass aloes, with the exception of *Aloe inconspicua*, grow and flower during the period of greatest rainfall. The flowering season peaks in early February, overall. This is towards the end of summer when the ground is moist and a large amount of precipitation can be expected.

Most larger species start to grow and come into flower before the rains, in the second half of winter, in spring and the early summer. The earliest species to flower, after grass fires, is usually *Aloe chortolirioides*. This may be in midwinter, June, and these early flowering plants are already in seed by late August. Some of the large species require warm spring weather to initiate leaf growth and bud development. The timing of flowering is usually earlier for species growing where the grassland is burnt in June and July. The ground receives direct sunlight and the warm weather starts off the growth cycle. Droughts and temperature variations affect the timing and abundance of flowering from one season to the next.

The smaller grass aloes characteristically flower amongst growing grass stems. Many of them produce long flowering stems in relationship to the size of their basal stem and root area. Their flowers are elevated slightly above the surrounding grasses and are conspicuous on account of their colour or shape against a uniform background.

The flowering performance of many species tapers off in grasslands that escape burning for a few years in succession. Abundant flowering is generally resumed during the first two seasons after grass fires.

The grass aloes that flower in the late winter and early spring release their seeds into the environment just before or at the beginning of the rainy season. The early summer-flowering plants seed mostly from December to early February. The late summer- and early autumn-flowering species set their seeds at the close of the rainy season as day length starts to decrease and temperatures begin to drop.

Seeds are dispersed mostly by wind and rainfall and also to some extent by insects such as ants. Germination results are always at their best after fire. The habit of restricting mass flowering to periods after fire which most grass aloes have, ensures not only the best chances for pollination, but also enhanced conditions for the germination of seed. In addition, the open habitat provides optimum conditions for the development of seedlings.

Grass aloe seeds, like those of so many South African plants, display the characteristics of delayed germination. This is particularly marked in some of the dwarf species that seed in early autumn. The first opportunity to germinate that most of these seeds have, is at the beginning of the following rainy season, usually eight or nine months later.

The fire-blackened stems of *Aloe hlangapies* plants growing in a firebreak on the perimeter of a timber plantation between Ermelo and Piet Retief in Mpumalanga. Several plants in this group were already in flower in October 2004.

Pollination

Grass aloes are pollinated by sunbirds (genus *Nectarinia*) and a variety of flying insects, including beetles. The most common insects observed pollinating many of the larger grass aloe species, are honeybees (*Apis mellifera*). In addition, ants may sometimes be pollinators of grass aloe flowers since various ant species are always observed on the blooms as well as inside them. However, some believe that ants visit flowers merely to 'rob' nectar and to collect pollen. Perhaps they also provide, by their presence, some protection to the flowers as they are quick to remove any larvae of other insects that may cause damage. Indications are that the formic acid associated with ants are damaging to pollen and that any grains sticking to their bodies quickly loose viability. Ants furthermore forage in a rather limited area and are therefore not very effective agents to carry pollen great distances between plant populations to effect cross pollination.

Detailed research is required by entomologists in association with botanists to determine which insect species engage in the most effective pollination of grass aloe flowers. Grass aloes flowering in relatively pristine grassland are visited by numerous insects, only some of which are likely to be pollinators. Grass aloes that inhabit regions with extensive stands of alien timber plantations have the fewest insect and avian pollinators visiting them at flowering time. This is one of the most significant factors projected to lead to the demise of grass aloes in these areas.

Above: A black sunbird feeding on the nectar of an *Aloe woolliana* flower. These sunbirds periodically visit populations of this species along the escarpment cliffs of the Drakensberg near Lydenburg in Mpumalanga. Black sunbirds are the most regularly encountered avian pollinators of grass aloes and are amongst the few species that visit grass aloes in flower in areas heavily planted up with alien timber. They either pierce the base of the flower to source nectar, as depicted here, or insert their beaks down the full length of the floral tube.

Sunbirds, particularly the black sunbird (*Nectarinia amethystina*), the greater double-collared sunbird (*Nectarinia afra*), the lesser double-collared sunbird (*Nectarinia chalybea*) and the malachite sunbird (*Nectarinia famosa*), regularly visit grass aloe flowers at the time of nectar flow.

Sunbirds are rare in areas where many of the grass aloes occur today, particularly where there are alien timber plantations and extensive crop farming. They are only partial to mountainous grassland where there are stands of proteas and bushclumps or to ravines with indigenous trees and shrubs. These birds are attracted by mass flowering of grass aloes. They are rapid feeders, moving quickly from one flowering aloe to the next. The typical feeding habits are for a pair of sunbirds or a small group of them to move across a large population of aloes, rarely returning to those they have already visited in the same feeding episode. The situation is different, however, when there are large groups of grass aloes in flower in a confined area. In such situations the birds regularly revisit the same flowers in a single feeding episode.

A typical example of the feeding behaviour of black sunbirds was observed in early January 2004. Good numbers of *Aloe boylei* and *A. kraussii* in a section of grassland near Hermannsburg in KwaZulu-Natal were in full flower. This patch of habitat is surrounded by timber plantations but is joined to other hilly grassland by an unforested corridor running along a river. The sunbirds spent about three hours visiting the flowers of both *Aloe* species. They worked systematically through patches of aloes. The feeding method the birds adopted was to visit some flowers on one stem and then some on other stems, not necessarily feeding on flowers on all the stems from the same plant. The sunbirds were more thorough when feeding on plants with the fullest nectar flow. The same aloes were repeatedly visited from about noon until the midafternoon.

One pair of sunbirds can potentially pollinate a huge number of grass aloe flowers in a short space of time. Insects are slower pollinators and their feeding is more readily disturbed by windy weather. Successful pollination by insects is usually attributable to large numbers of them visiting a flowering population of grass aloes.

Several populations of *Aloe kniphofioides* near Piet Retief in Mpumalanga used to be pollinated by black sunbirds and malachite sunbirds. The area was heavily forested with pines but with extensive river-bank corridors of grassland, similar to those near Hermannsburg. In recent times new plantations have been planted right down to the banks of the streams. Only rocky areas with shallow soil remain, unsuitable habitat for pines. The remaining *A. kniphofioides*, growing in isolated groups, are rarely if ever visited by sunbirds at flowering time. Very little seed is set nowadays and this seems to be at least partially attributable to the lack of attractive scattered groups of flowering aloes that used to attract pollinating sunbirds.

Honeybees are particularly common pollinators of *Aloe ecklonis* on the Highveld and in various cattle- and maize-farming areas of KwaZulu-Natal and the Free State. The increased seed production in many *A. ecklonis* populations over the years is related to the greater number of honeybees. Certain crops require honeybee pollination and apiarists move their hives to some areas at the time when grass aloes are in flower. In addition, the frequency of buildings in formerly featureless grassland affords greater opportunities for swarms to establish their hives.

Suburban gardens in cities often carry high numbers of honeybees. This artificial habitat is significant in attracting honeybees to areas where many of the range of natural grass aloe pollinators are no longer present. It is likely that honeybees have, in many places, become the predominant pollinators of grass aloes. This does not imply that they are more efficient at pollinating the flowers than a wider range of less dominant but equally active insect pollinators.

Some interesting observations were made on the pollinators of *Aloe woolliana* on the Drakensberg escarpment near Lydenburg and on the Steenkampsberg west of the town, in the spring of 2004 during the peak of the flowering season. Both areas have large populations of *A. woolliana*. Plants on the Steenkampsberg were repeatedly visited by at least eight species of potential pollinating insects during a three- to four-hour observation period. This was during late morning and early afternoon when pollinators are most active, in warm, sunny weather. Near Lydenburg no insect pollinators were observed on *A. woolliana* flowers. The aloes here were growing on an extensive broken, grassy escarpment cliff face surrounded by plantations on all sides. A few of the range of insects observed on the Steenkampsberg were also feeding here in large numbers, but on flowers of *Greyia radlkoferi*. These flowers were producing copious nectar at the time and as a result of the strength of the nectar flow, were attracting most of the potential aloe pollinators.

Pollination by insects and sunbirds is affected by mist and cloud cover and also by windy weather. In the case of species flowering in the first half of summer, windy weather is frequent in the late morning and early afternoon, the time when pollinators are at their most active. Autumn-flowering species flower at a time when a large amount of mist and rain can be expected. During seasons of prolonged inclement weather the pollination of several species is very poor. This applies particularly to the very rare newly described *Aloe craibii*. These plants grow on rocky, grassy hillsides near Barberton in Mpumalanga. The few known colonies are found in areas heavily afforested with timber plantations, a form of land use that excludes many pollinators. Misty weather at flowering time exacerbates the pollination problem and in some years few, if any, plants produce seed.

Dwarf grass aloe species set good amounts of seed when much of their habitat is surrounded by suburbia. The most interesting example of this is *Aloe parviflora* near Botha's Hill north of Durban in KwaZulu-Natal. Colonies of plants are usually encountered in rocky, hilly grassland on the outskirts of suburbia. The seed set is usually very good from one year to the next, with a healthy recruitment of young plants to the population. Numerous visits have been paid to the habitat at flowering time, but pollinators have not been observed.

Aloe inconspicua grows in savanna grassland near Estcourt in KwaZulu-Natal. The limited area where it is found is used for cattle ranching but none of the ecological processes present in the grassland have been significantly disturbed. Consequently the plants successfully attract pollinators each year, resulting in healthy populations containing some seedlings, evidence of healthy regeneration.

It seems likely that most dwarf grass aloes have specific pollinators owing to their specialised floral structures. Additional research data concerning this subject are required. Many days of field observation on most species at the peak of the flowering season failed to yield observations on pollinators.

Development and dispersal of seed

A number of *Aloe* species that do not fall within the grass aloe group produce enormous amounts of seed for liberation into the environment. The results are that some species become common and dominant in certain habitats. This is particularly evident where some of the biological processes in an environment have been modified. Heavy over-grazing in parts of Sekhukhuneland in Limpopo Province has led to a proliferation of *A. globuligemma* and *A. cryptopoda*. In the grasslands around Johannesburg in Gauteng where regular grazing and trampling of the veld no longer occurs, *A. greatheadii* var. *davyana* is becoming abundant.

There are several mechanisms that control the numbers of aloe seeds available for germination and that kill off newly germinated seedlings. These processes form the subject of this chapter.

The seed chalcids of the family Eurytomidae are small, prolific wasps whose larvae feed on developing seeds. At least one species, *Bruchophagus aloineae*, feeds specifically on developing aloe seeds. There may also well be some undescribed seed chalcids that feed on grass aloe seeds. The presence of the wasps is evident from small round holes bored into the seed capsules as well as the individual seeds. Huge quantities of aloe seeds can be destroyed by these wasps. They occur very widely in many aloe habitats in South Africa and are active throughout the year. There are no species of grass aloes in South Africa that are exempt from the attention of the wasps.

Levels of seed parasitism vary greatly from one flowering season to the next. In areas where aloes have not been reduced in numbers through habitat destruction and alteration, the wasps limit the amount of seeds liberated into the environment. Problems arise, however, with many of the grass aloe species. In regions afforested with alien trees and heavily grazed or in degraded grassland, the wasps often have a negative effect. High levels of parasitism mean that very little seed is available for dispersal. Many grass aloe species already suffer from lack of effective pollination owing to changed habitats. In such cases the seed-eating wasps exacerbate the decline of grass aloes. The largest grass aloe species, which produce the most seeds, are the most attractive to the wasps and the smaller species, which produce limited amounts of seeds, the least attractive.

In addition to parasitism by the seed wasps, grass aloe seeds are eaten by beetle larvae. The grubs spin cocoons around the ripening aloe seed capsules. They usually eat all the seeds ripening on the stem and even if they do not, their droppings attract fungi which generally destroy most remaining seeds.

The effects of the wasps on *Aloe minima* var. *minima* and *A. minima* var. *blyderivierensis* are a good example of wasp parasitism on smaller grass aloe species. Around The Brook in Mpumalanga, where there are still some large areas of hilly grassland, *A. minima* occurs mostly as single plants widely scattered across the veld. Some seeds are parasitised often where half a dozen or more plants grow close to one another. Single, isolated plants may escape the attention of the wasps altogether. A colony of *A. minima* var. *blyderivierensis* in an area heavily planted-up with alien trees near Graskop in Mpumalanga, is heavily parasitised by the wasps in some years. The plants would certainly be less parasitised by the wasps were they widely scattered across the grassland and not concentrated in small populations by plantations.

In some years when *Aloe chortolirioides* flowers in the mountains near Barberton in Mpumalanga, ripening seeds are eaten by the wasps and beetle larvae. Pollination is very poor in the heavily forested areas owing to a lack of pollinators. The limited seed set has often largely been destroyed by the insects over the last two decades of field observations.

It is rare to find any grass aloe species escaping some level of seed parasitism in a given flowering season. Where grass aloes flower close to other prolific aloes flowering at the same time such as *Aloe arborescens* and *A. spicata*, the wasps tend to parasitise the more prolific species. The beetle larvae, however, can be found on any ripening aloe seeds and are often present without the wasps. A range of lepidopterous larvae eat ripening grass aloe seed capsules and the enclosed seeds. They are not specific to grass aloes but eat them when they have depleted other foliage on surrounding plants such as species of *Albuca*.

Grass aloe seeds are best dispersed in dry, windy weather. The seed capsules dry out thoroughly and split open lengthwise. Cool, cloudy weather prevents the seed capsules from receiving enough sunlight to split open properly. If such weather is accompanied by rain at the time the capsules start to split, many seeds are destroyed by fungi.

Wind, water runoff after rain and insects, particularly ants, are responsible for dispersing the seeds. Wind is usually the main dispersal agent and has the important function of dispersing seeds widely in the environment. Grass aloe seeds, like those of most aloes, are prone to destruction by damping-off fungi when they germinate. Seeds that are widely and thinly dispersed by wind are least affected by damping off. Windy weather ensures that seeds do not accumulate in large concentrations around the parent plants.

Much of the interior of South Africa is subject to periodic severe summer hailstorms. Hail destroys nearly all the ripening grass aloe seeds in some years. The species most affected are those that flower and form seed from October to December when the incidence of hailstorms is usually most likely. Hailstones knock off most ripening seed capsules and damage those that remain. The seed in the damaged capsules is usually destroyed by fungi. Hailstorms are particularly destructive in some areas such as the Piet Retief District of Mpumalanga.

The factors that control grass aloe seed production are very variable. Good pollination may or may not coincide with high levels of seed parasitism. Inclement weather sometimes lead to high levels of fungal attacks in years when seeds are not parasitised. The various combinations of factors that destroy grass aloe seeds ensure that most species have strong seed sets only a few times each decade.

Road and railway-line reserves

Road and railway-line reserves provide many South African plants with small areas of well-preserved habitats. These reserves consist of fenced off strips of land on either side of a railway line or road. In the case of roads, the fences generally mark farm boundaries whereas railway lines are fenced off not only from farms but also from any roads that run parallel to them. These reserves consist of habitat isolated from the surrounding veld. They escape extreme grazing pressures and regular trampling by livestock such as cattle and sheep.

A number of grass aloe species are conserved in these reserves and some of them become common in these situations. The reserves themselves are usually rather narrow but they have the beneficial effect of preserving some of the ecological processes across a wide range of habitat niches in a given area.

Grass growing in road reserves in flat areas is usually cut in the winter and the hay baled for livestock fodder. This habitat is therefore generally cleared of tall grasses. The clear habitat results in good successive grass aloe flowering seasons for several years in succession. These seasons of abundant flowering are broken only by periodic droughts. Some grass aloe species such as *Aloe ecklonis* establish large populations under these conditions.

In some parts, particularly the Albany District of the Eastern Cape, grass aloes are found in significant numbers only in road and railway-line reserves. In other areas such as the grassland around Leandra on the Highveld, *Aloe ecklonis* is preserved in some very large populations in a railway-line reserve, extending over several kilometres. The surrounding grassland on farms is heavily grazed and the *A. ecklonis* populations there are small and widely scattered.

Grass aloes have disappeared from many parts of South Africa that are heavily populated and over-grazed and where there are no fenced road reserves. The reserves provide a significant opportunity to examine grass aloe populations where certain contemporary pressures on the veld such as over-grazing are largely eliminated. The reserves remain part of a fire ecology and are burnt as regularly as the surrounding grassland. The exceptions to this situation are grasslands in flat areas and crop-producing regions. Grassland here is usually mowed mechanically and the hay is baled. The mowing provides not only firebreaks for crops but also a valuable source of additional fodder. Some grass aloes, particularly *Aloe boylei, A. ecklonis* and *A. kraussii*, have responded favourably to these artificial conditions.

The significance of road reserves for the preservation of flora has been formally recognised in South African flora conservation legislation. These reserves are, however, currently subject to various forms of habitat degradation in some regions where grass aloes occur. Railway-line reserves, too, are falling into disrepair. This is mostly related to a decrease in the significance of railway transport in recent times. Some road and railway-line reserves harbouring grass aloe populations are discussed below, with specific reference to changes in their status as former corridors of plant conservation.

One of the older railway lines in South Africa is the one built by the Cape Government between Grahamstown and Alicedale in the Eastern Cape. The line was completed in 1879. The reserve area adjacent to the railway line

Above: The railway-line reserve near Coldspring in the Eastern Cape is an important sanctuary for *Aloe myriacantha*. These plants have all but disappeared from the Coldspring area owing to grazing by cattle and sheep which trample the grassland. Several aloes were found in the railway-line reserve but not in the adjacent heavily grazed veld.

conserved a large transect of the veld that characterises the grassland of the Albany District. The secondary road from Grahamstown to Alicedale runs alongside the railway line. Most of the area between the road and the railway forms a large reserve varying in width according to topography.

This reserve has served as an important area for the preservation of *Aloe myriacantha*. This species gradually disappears from areas heavily grazed by sheep. Sheep farming is practised throughout the Albany District, which used to be a stronghold of this grass aloe in South Africa. A brief survey was done just before the peak of the flowering period in late February 2004. The survey attempted to correlate railway-line reserve populations of *A. myriacantha* to the status of plants in the surrounding veld stocked with cattle and sheep. No aloes were found in well-grazed farmland adjacent to populations of these plants in the reserve areas.

The reserves themselves are, however, undergoing transformation. Part of the road and railway line reserve near Coldspring, with a hill inhabited by *Aloe myriacantha*, is now used for grazing stock. In the area between Coldspring and Highlands the fencing has fallen into disrepair and local inhabitants also use the reserve grassland for grazing stock. These problems are compounded by weeds springing up under over-grazed conditions.

Road and railway-line reserves are an important source of seeds for repopulating the surrounding veld with new plants. This is particularly so where land use patterns change, for example a transition from sheep farming to game farming, with different types of grazing. Degradation of grass aloe populations in the reserves can be expected to contribute significantly towards the local extinction of some grass aloe species such as *Aloe myriacantha*.

The Leandra area on the Highveld in Mpumalanga has for many years been home to large road and railway-line reserve populations of *Aloe ecklonis*. Over the last decade several extensive informal settlements have been established in the area. These reserves are currently undergoing changes to their former status as conservation corridors of highveld grassland. The countryside carries herds of sheep and cattle and is also used for seasonal crop production. The aloes are common only in grassland that is no longer used for grazing, or on the immediate outskirts of some of the towns, where informal settlements have not developed.

The fencing around railway-line reserves close to informal settlements has largely been removed by people living there. Livestock kept in these settlements graze the veld in the reserves. Most flower stems get grazed and trampled by livestock and they are also periodically snapped off larger plants by the hooves of the grazing animals. Most of the once numerous younger plants have disappeared and flowering and seed set are negligible in comparison with former times.

The road and railway-line reserves adjacent to farmland have fared better. Some farmers graze their livestock in the reserves but this is occasional. The grassland in the level reserves is cut towards the late autumn and the hay collected and baled for livestock fodder. Many aloes get cut back by the machinery, resulting in most of these plants developing a clump-forming habit with numerous stems. This has led to a huge increase in numbers of flowering stems per plant, and to far more flowers in comparison with less disturbed populations. There is also an enormous seed set during years of good rainfall.

There are some very large colonies in railway-line reserves where the fencing remains intact. The only pressures exerted on these populations are occasional grazing and unseasonal late summer or autumn fires during droughts. The veld adjacent to these reserves in many cases has few or no aloes owing to grazing pressures. Several significant bulbous plants have virtually died out from this area in recent times, the most significant species being the rare *Nerine gracilis*. *Cyrtanthus stenanthus* and *Haemanthus montanus* have suffered a similar fate. The reserves play a critical role in preserving these plants and *Aloe ecklonis* in this grassland region. The bright yellow *A. ecklonis* flowers attract a range of mostly insect pollinators, particularly honeybees and beetles. This in turn leads to a good seed set, with numerous seeds germinating and giving rise to many young plants.

Aloe dominella owes its continued existence in several parts of KwaZulu-Natal to road reserves. This species declines sharply in cattle ranching areas without rocky outcrops. Plants are readily damaged by grazing animals, particularly cattle. Plants wedged between rocks or around boulders are less easily damaged, but this habitat represents only a small portion of the overall area utilised by the species.

Northeast of Mooi River there are significant road reserve populations of *Aloe dominella*. In the adjacent hilly grassland used for cattle ranching the effects of grazing are clear. The plants are now found largely on very steep hills, and not in the flatter grassland where they were once widespread. They are, however, well represented in road reserves with flatter or gently undulating ground.

Near Rorke's Drift in the Dundee area of KwaZulu-Natal, *Aloe dominella* is, in some parts, only a road reserve plant. The species has disappeared from flatter grassland with occasional boulders and sheets of exposed rock.

Plans are now being implemented for replacing cattle farms with huge game ranches. This initiative has been brought about as a result of continual stock theft. It is possible that *Aloe dominella* will recolonise grassland areas where it was once more numerous as a result of different types of grazing. In such cases, road reserve populations of this aloe can be expected to provide an important source of seeds for repopulating the adjacent grasslands.

Near Piet Retief in Mpumalanga, road reserves have an interesting modern function. They preserve *Aloe hlangapies* in areas heavily afforested with eucalypts and pines. They also foster opportunities for the hybridisation of *A. hlangapies* and *A. ecklonis*, where these two plants grow close together. The surrounding land is heavily afforested with alien trees, implying that the pollinators that feed on the two species visit the same flowers over and over. In several cases where the flowering season of two species overlap, most plants in the populations are hybrids. Northeast of Piet Retief in open, hilly grassland that has escaped alien afforestation, the typical *A. hlangapies* is still regularly encountered. Populations where the two species hybridise are mainly the result of changes to biological processes, brought about by the establishment of extensive plantations of alien trees.

Road and railway-line reserves are important for the preservation of grass aloes in the modern South African grassland environment. These reserves do, however, have an artificial man-made status that regularly changes. Road and railway-line populations of grass aloes, although of much significance, are unlikely to be secure in the long term.

The plants

1. *Aloe albida*

Aloe albida (Stapf) Reynolds

This dwarf species is found in misty mountainous grassland in Mpumalanga, between Barberton and the border of northern Swaziland. The plants also occur in neighbouring parts of Swaziland. This region is densely forested with timber, chiefly eucalypts and pines. In areas away from the plantations many parts of its habitat are being transformed by black wattle (*Acacia mearnsii*), an introduced species from southeastern Australia and Tasmania. *Aloe albida* occurs in the mist belt mostly at an altitude of 1 600–1 800 m and in several areas it grows alongside *A. chortolirioides* and the recently described *A. craibii*.

The plants have declined greatly in numbers over the last 20 years as a result of habitat loss to alien timber plantations and transformation of the grassland by black wattles. The lowering of the water table through extensive afforestation and invasion of the grassland by annual and perennial weeds is another deep-seated problem in the heavily afforested areas. Few large colonies remain and in places where the species used to be well represented, numbers have declined sharply since the 1980s. The plants continue to decline as a result of recent extensions to pine plantations.

The habitats occupied by *Aloe albida* are very similar to those occupied by *A. saundersiae*, a diminutive species from central KwaZulu-Natal. *A. albida*, however, is able to grow on the north- and northwest-facing topmost mountain slopes owing to prolonged misty and cloudy weather in summer.

Prior to afforestation, *Aloe albida* used to occur on rocky outcrops, in short mountain summit grassland and on broken cliff faces covered in short grass and moss. In most of the habitat currently occupied by the plants, only one or occasionally two of these three niches are available to the plants. *A. albida* is usually most common in mountain summit grassland and more scattered and rarer in the other two types of habitat. Most of the short mountain summit grassland away from extensive rocky areas is now covered in alien timber plantations.

Three populations of *Aloe albida* growing on mountain summits between Barberton and the Swaziland border have been studied over a period of some 20 years. The best-known colony is the one on the top of the Saddleback Mountains behind Barberton. This is also the type locality.

Colony on the Saddleback Mountains behind Barberton

A colony of *Aloe albida* occurs on the top of the Saddleback Mountains, at the northerly end of its distribution range. An important historical photograph of what this habitat used to look like prior to the advent of afforestation is to be found in Reynolds's (1969: 113) *Aloes of South Africa*. Most of this area has now been planted with pines. The ridges around the type locality have escaped afforestation as they are too rocky for alien trees to take root.

The aloes grow on the higher ridges that attract most of the mist. The habitat is very rocky, with pockets of soil on

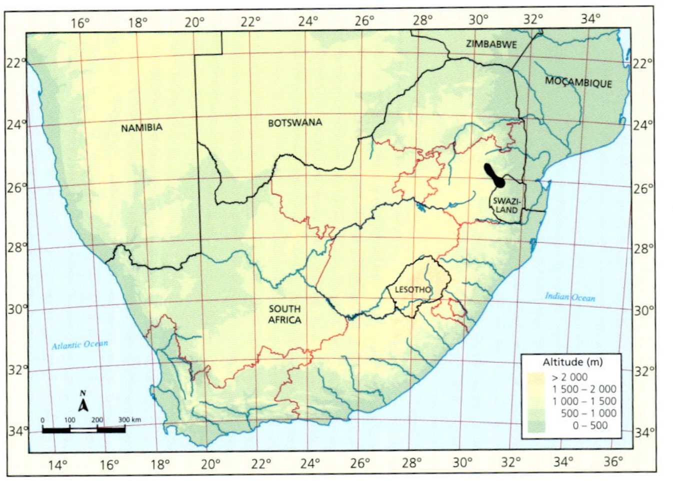

Distribution of *Aloe albida*

the broad ledges of small cliffs and between boulders on the summit. The ridges face north and many of the *Aloe albida* plants grow in these sunny positions. They are found mostly on those sections of the ridges where mist condenses and falls as soft rain. The remainder of the habitat is too dry for them. Areas where the mist swirls up the Saddleback Mountains during cloudy weather are good indications of sites where the plants can thrive are. Some peaks, not necessarily the highest, attract more precipitation from mist than others. The Saddleback Mountains is on the perimeter of the mist belt and the lower grassy slopes, which are otherwise ideally suited to the requirements of these plants, do not attract enough mist. The cliffs and boulders on top remain moist for long periods after misty or rainy weather and the aloes are irrigated from seepages even in the thinnest soil. Thundershowers are common in summer and heavy falls are usually followed by a few hours of mist.

The aloes occupy much of the habitat and some of them have even taken root in clumps of moss. *Aloe chortolirioides* grows mostly on the more level sections of the ridges and is often found only a few metres away from *A. albida*. *Brachystelma swazicum* is commonly found in the same pockets of soil as the clumps of *A. albida*.

Populations on the mountains between Barberton and the Swaziland border

Aloe albida used to be plentiful on several mountain peaks and lower hills between Barberton and the Swaziland border. Most populations have declined markedly over the last 20 years. One colony used to throng the short grassland on a mountaintop at an altitude of about 1 700 m. There used to be thousands of plants growing in all the sites occupied by this species where it occurs in extensive habitat.

The aloes grow in short grassland, in pockets of soil amongst seasonally damp mossy rocks and more rarely amongst moss and short grass at the edges of sheets of exposed rock. The plants were at their most abundant on

Right: ***Aloe albida*** growing high up in the mist belt close to the type locality near Barberton in Mpumalanga. It flowers in early autumn when the grass blades are withering.

the mountain summit, with scattered groups on the lower slopes. They occurred mainly on east-, west- and southwest-facing slopes, avoiding the arid northern aspects of the lower mountain slopes. *Euphorbia pulvinata* is common on the northern slopes and where this succulent occurs, *Aloe albida* is usually absent.

Seed set used to be good when there was dry weather at the flowering time, enabling pollinators to visit the flowers. The area is surrounded by alien tree plantations, but the adjacent region to the east consists of unforested rocky mountaintops not suited to timber production. This limited amount of mountaintop habitat, several kilometres long, was sufficient to sustain most of the biological processes necessary for the survival of the plants.

Over the last two decades, timber plantations have been extended over much of the former mountain slope habitat. Invasive annual and perennial weeds are frequent, degrading the grassland in the process. *Aloe albida* is now mainly an inhabitant of rocky outcrops on the mountain. This habitat used to be the least attractive to the plants but currently represents the last suitable niche where the plants can continue to grow.

The continued existence of *Aloe albida* in the lower lying areas below the mountain summits is even more precarious. Most of the plants grow on rocky outcrops beside forestry roads, in remaining strips of grassland that amount to only a few hundred square metres in extent. Logs are often dropped on this habitat during tree felling operations and the bark and branches that remain once the logs have been transported, have greatly increased the quantity of fungi present in the veld. This has negative effects on several grassland plants, including *A. albida*. The soil disturbance that occurs during logging operations favours the establishment of weeds, which further degrades the habitat. It is likely that most of the small *A. albida* colonies that still occur in roadside grassland, will disappear in the coming decades.

Growth cycle and the role of fire

Aloe albida is dependent on fire to clear moribund grass from the mountain summit grassland, preferred habitat of this species. Plants growing in moss and at the edges of sheets of exposed rock are less thoroughly burnt in grass fires and then only when the sparse grass cover is ignited by strong winds that drive the fire. The recent contraction of many *A. albida* populations to rocky outcrops is partly related to the infrequency of fires in nearly all the remaining habitat.

The plants starts to sprout leaves late in November or occasionally October in years when the summer rains begin early. They grow mostly from late November to late January. Flowering may begin as early as the second half of February but March is usually the peak of the flowering season, which extends until the second half of April. The timing of flowering is related to onset of the summer rains. When the rains are late, or during droughts when the plants rely on mist for moisture, the peak of the flowering season is often the first two weeks of April.

Status of wild populations

Aloe albida has suffered two major population declines. The first was the destruction of the greater part of its habitat through afforestation. The second occurred when many of the remaining populations that had stabilised after initial afforestation were subject to additional pressures, mostly further afforestation, invasion of the habitat by weeds and logging operations.

An effort could be made to locate further *Aloe albida* populations in the mist belt mountains northeast of the Swaziland border. Much of this area is not afforested with aliens and may yield some significant populations. The Songimvelo Nature Reserve northwest of the Swaziland border may well harbour significant populations of *A. albida* and the species has already been recorded in the nearby Malolotja Nature Reserve in the adjacent section of Swaziland.

Aloe albida is one of the dwarf grass aloe species that are successfully pollinated in corridors of mountain grassland between timber plantations. Some of the remaining populations could, therefore, be managed in conjunction with timber companies.

Cultivation, propagation and horticultural potential

Aloe albida makes for an attractive dwarf container plant. It thrives in fine soil to which a small amount of sieved, sandy soil and sieved, very well-rotted leaf mould have been added. Plants grow well in terracotta pots kept in lightly dappled shade or morning-only sunlight. They require plenty of free air circulation to grow and flower well and should also be kept well watered throughout the growing season. The aloes should be lightly and occasionally watered in the winter months during spells of warm weather. Root mealybug is a problem, particularly if the containers in which the plants are growing overheat in summer.

The species is best propagated from suckers or seeds. Large plants produce a few suckers from time to time and these can be rooted in fine silt. Seed should be sown in spring so that it has the chance to germinate and the seedlings have the chance to establish themselves before the heat of summer sets in. Seeds should be lightly and evenly scattered across the surface of the germination medium and afterwards just covered with the same mixture. A good germination mix is the silty soil described for the cultivation of adult plants.

Germination takes two to four weeks on average and not all seed germinates in the first year after it has been sown. The seed trays should be kept moist after sowing. Seedlings should be well watered after germination, but care has to be taken to ensure that they are not kept too wet as it may lead to damping off. Most damping off problems can be avoided by sowing the seeds during warm spring weather. The seedlings are then established by midsummer when attacks by damping-off fungi are at their most troublesome. *Aloe albida* produces seed in cultivation but requires hand pollination to produce a good seed set.

Aloe albida appears to be one of the dwarf aloe species that can be successfully multiplied by the techniques of tissue culture.

2. *Aloe boylei*

Aloe boylei Baker

This robust grass aloe usually has very broad leaves. In some areas the flowers and shape of the racemes are virtually indistinguishable from those of *Aloe cooperi*. This is particularly evident in the Greytown area of KwaZulu-Natal.

Aloe boylei was reduced to synonymy by Glen & Hardy (2000: 19). The difficulties that arise from this decision are discussed in the chapters on *A. ecklonis* and *A. kraussii*. Glen & Hardy do not give reasons for their inclusion of *A. boylei* under *A. ecklonis*. This tends to leave a vacuum in the construction of botanical knowledge and it may well be that if *A. boylei* has close affinities with other grass aloes, these lie elsewhere. Reynolds (1969: 154) and Bornman & Hardy (1971: 45) comment on the close similarity of the flowers and shape of the inflorescences when *A. boylei* is compared to *A. cooperi*. The leaves of the two species are cited as the main distinguishing features. This distinction is not always constant, however, and some thin-leaved plants of *A. boylei* are close to what may currently be regarded as *A. cooperi* without keeled leaves.

The species concept of *Aloe boylei* would merit reassessment. For this evaluation to add authenticity to the construction of knowledge about grass aloes, detailed fieldwork would seem to be an indispensable component of the research.

Aloe boylei occurs in the grasslands of the eastern Cape, KwaZulu-Natal and Mpumalanga. Its distribution range also enters Limpopo Province in the Magoebaskloof area. Plants growing in the grassland of the interior, where the bulk of the populations are found, usually occur in mist belt areas. *A. boylei* has a preference for rocky grassland and the plants often form compact groups in this type of habitat.

The species was studied at two localities near Greytown in KwaZulu-Natal. One of these lies well within the mist belt. The other is situated right at the edge of the mist belt. Conditions become drier only a few kilometres away as the altitude decreases. *Aloe boylei* was not found in the more arid grassland that typifies conditions closer to Muden, a small settlement northwest of Greytown.

Occurrence northwest of Greytown

Aloe boylei occurs northwest of Greytown in hilly grassland, most of the area densely afforested with eucalypt and pine plantations. Some of the hillside grassland and most areas near streams have been left intact. Strips of grassland are sometimes several hundred metres wide near these streams if the area is rocky and unsuitable for planting timber. These corridors of habitat along streams usually link up with other similar areas including large stretches of hillside grassland.

The plants are plentiful in the rocky areas with short grass. Many seedlings and young aloes in all stages of development are found in places with the shortest grass. The veld was not grazed by livestock during the period the data were collected. The lack of trampling by livestock provided conditions suitable for the optimum development of young plants. The aloes were most frequent in the vicinity of seepage areas and around sheets of exposed rock.

Pollination of the aloes was very favourable, ensuring a

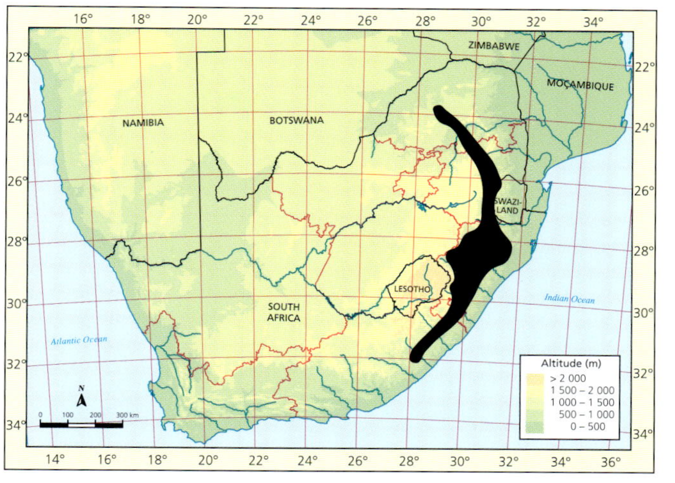

Distribution of *Aloe boylei*

good seed set. The plants had also established themselves in good numbers in the nearby road reserve. They were particularly plentiful on an artificial road embankment, a situation with less plant cover and thinner grass than the surrounding veld.

Near Hermannsburg in the Greytown area

Near Hermannsburg, *Aloe boylei* occurs in a region densely utilised for timber production. Patches of open grassland have been left intact, mostly along watercourses. One of these areas of grassland has been preserved through the initiatives of local people. This grassland, which covers a gently sloping west-facing hill, contains most of the range of grassland flora found in the Greytown area.

The more elevated parts of the grassland contain large numbers of *Aloe kraussii* whereas *A. boylei* is confined to rocky lower lying areas along a stream. The plants of *A. boylei* come into flower in the second half of December and are usually in seed by the beginning of February. Seedlings and young plants are not as common here as they are at the locality northwest of Greytown. Much of the seed goes to waste amongst the rocks and the remainder falls into tall grasses. These situations are not conducive to the development of young plants. The plants are, however, found in tightly packed groups in the limited amount of short grassland, adjacent to some of the boulder outcrops.

The flowers are regularly pollinated by black sunbirds (*Nectarinia amethystine*) and honeybees (*Apis mellifera*). The presence of regular pollinators is attributable to the abundance of nectar and pollen-bearing flowers in the grassland reserve, linked to other areas of grassland by riverine corridors.

Patches of grassland in timber-producing areas are often used for grazing livestock. The few grass aloe species that do manage to flower in places where they are not trampled, are usually insufficient to attract pollinators. The concept of grassland reserves that exclude grazing livestock, provides a significant method for preservation of biological processes in areas heavily planted with timber. This may in due course amount to the only viable method for the *in situ* conservation of most grass aloe species that occur in timber-producing areas that still have riverine grassland corridors.

Growth cycle and the role of fire

Aloe boylei produces leaves in October and November, at the beginning of the main summer rains. Leaf development is usually completed by the end of November and the plants start to form buds early in December. The peak of the flowering season is in late December and early January. Plants are sometimes found in flower until the end of January in years when the summer rains are late.

Seeds are liberated in February and are scattered about the grassland environment by wind and water runoff after rain. Some of the seeds germinate in late summer and early autumn. The bulk of the seed set that remains in the environment germinates once the main rains begin the following summer. *Aloe boylei* is a prolific seeder and this accounts for the abundance of the plants in the limited number of areas where grazing livestock are absent.

Fire plays an important role in keeping the grass cover short. Fires stimulate mass flowering in summers preceded by a winter grass fire. *Aloe boylei* stems thicken with dead leaf material during successive years without grass fires, attracting fungi and various pests such as mealybug and scale that are burnt off by grass fires. *A. boylei* occurs mostly in sections of the habitat with the sparsest grass cover. In these situations seedlings are least affected by annual fires lit to burn timber-protecting firebreaks.

Status of wild populations

The species concept of *Aloe boylei* may well be redefined in coming years and this will inevitably affect the status of the plants. As the species is currently understood, it ranks amongst South Africa's better represented grass aloe species.

Many areas of grassland where the aloes used to be common have been converted to timber plantations. This applies particularly to the Greytown area of KwaZulu-Natal and the Haenertsburg and Magoebaskloof areas of Limpopo Province. The remnants of many formerly extensive populations survive in road reserves or occasional limited areas set aside for the preservation of grassland plants.

Aloe boylei, like *A. ecklonis*, is a prolific seeder and capable of repopulating grassland areas where grazing livestock is moved elsewhere. Road reserve populations supply an important source of seeds for replenishing grassland where numbers of *A. boylei* have declined as a result of over-grazing.

Cultivation, propagation and horticultural potential

Aloe boylei is an easy plant to grow in cultivation. The aloes should be grown in large, deep containers, in deep soil in rockeries or on sloping ground not subject to soil disturbance. The plants have a deep root system that has to develop well to ensure regular flowering.

The cultivation mixture should be well drained. A useful soil mix consists of one third gravelly sand, one third sieved, well-rotted compost or kraal manure and one third sieved sandy soil. *Aloe boylei* thrives in well-ventilated positions. The plants may be exposed to day-long direct sunlight, sun for only half a day and even to lightly dappled shade. They grow and flower well under all these conditions.

The aloes should be well watered at regular intervals throughout the late spring and summer months. Watering should be substantially reduced as the leaves start to die back during April, in preparation for the winter dormancy. The plants should be kept dry during winter. They may be watered thoroughly a few times at the end of winter or in early spring in preparation for the summer growing season.

Aloe boylei is easily propagated from seed. Seeds may be sown in any well-drained seedling mix, provided that it does not contain large amounts of poorly rotted organic material. This encourages damping-off fungi, which rapidly destroy seedlings during the first few months after they have germinated. Seeds should be sown either in deep seed trays or pots. They should be evenly scattered over the surface of the soil and thinly covered with the germination mixture.

The containers in which the seeds have been sown should be kept moist throughout the germination period and for the remainder of the summer while the young plants are developing. Watering should be gradually reduced as the autumn approaches. The seedlings should be kept mostly dry during winter. They do, however, benefit from an occasional light watering on mild winter days.

Seedlings are ready for transplanting at the beginning of their third growing season. The species will even start to flower if it is left to grow in large seed trays. A plant produced from seeds sown late in February 1999 flowered for the first time in its seed tray in December 2004 and again about three weeks later in January 2005.

Aloe boylei can be struck from stem cuttings taken in late winter or early spring or in autumn. Cuttings should be snapped off the parent plant if possible. Alternatively, the cuttings should be taken with a sharp knife. These cuttings should be allowed to dry and heal for about 10 days and then rooted in sand, gravel or very well-rotted leaf compost. The cuttings are ready for transplanting once they have developed a root system. Plants rooted in autumn should be transplanted at the beginning of the following spring.

This is a showy grass aloe with striking flowers and unusual, broad leaves. A selection made in favour of plants with the largest racemes and broadest leaves would be useful for the landscaping industry. Mass plantings of this grass aloe are as effective as those of *Aloe cooperi*, another species that thrives under cultivated conditions.

Jürgen Greve of Hermannsburg sent some photographs of an *Aloe boylei* growing in his garden that had formed a plantlet on one of its flower stems, complete with a rosette of about 10 rosulate leaves. This had occurred spontaneously and was not recorded for any grass aloe species during the course of the research for this book. In such cases, the plantlet should be removed from the stem and rooted in moist sand. These plantlets develop vegetatively from modified buds, technically called bulbils or bulblets. The phenomenon, known as vivipary, is common in members of the century plant family (Agavaceae), but has only occasionally been noted in some aloes, especially in *A. bulbillifera*, a species from Madagascar.

Right: ***Aloe boylei*** in rocky grassland in a privately maintained veld reserve near Hermannsburg east of Greytown in KwaZulu-Natal. The leaves of these plants are particularly broad and robust.

3. *Aloe chortolirioides*

Aloe chortolirioides A.Berger

This aloe has a more limited distribution than *Aloe woolliana*, a very similar species. It is known from the mountains between Barberton in Mpumalanga and the Swaziland border. The plants are also found in the adjacent parts of Swaziland near the border with Mpumalanga.

Reynolds (1969: 127–128) described *Aloe chortolirioides* var. *boastii*. These plants are very similar to the typical *A. chortolirioides* and do not vary at all with respect to their autecology. The varietal distinction is therefore not upheld in this work. This should not, however, be seen as invalidating the concept Reynolds used to circumscribe variety boastii. The habitat currently occupied by *A. chortolirioides* has shrunk to such an extent through afforestation, that the full variation across populations of plants that Reynolds observed no longer manifests itself. Most colonies consist of small groups of aloes hemmed in by pine plantations on all sides.

Aloe chortolirioides occurs on mountain summits as well as north-, south-, east- and west-facing slopes. The plants grow in rocky grassland, but usually not on cliffs even if there is grassy areas on some of the ledges. This is in contrast to *A. woolliana*, which is partial to rocky, grassy cliff faces. The mountainous habitat where *A. chortolirioides* grows is subject to prolonged and frequent misty weather in summer and autumn, from November to March.

The flowering season has usually been completed by late September, about four to six weeks before the main summer rains begin. *Aloe chortolirioides* is completely dependent on fire cycles for flowering, and changes in the distribution and frequency of fires negatively affect the status of the species.

Some representative colonies of *Aloe chortolirioides* in the mountainous grassland near Barberton are discussed in the text that follows.

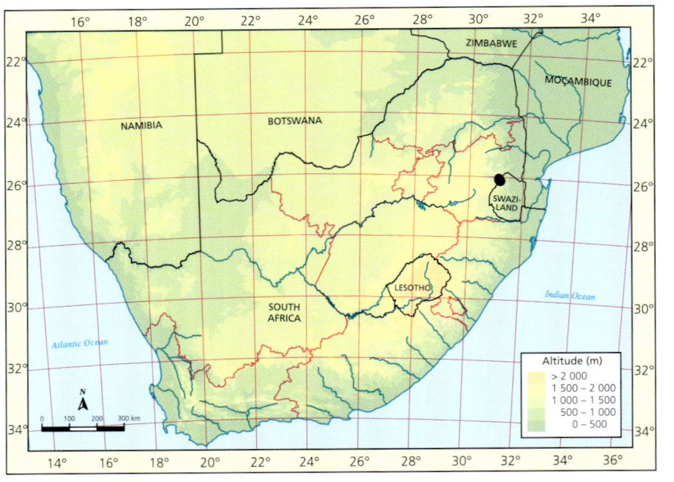

Distribution of *Aloe chortolirioides*

Colony on the Saddleback Mountains near Barberton in Mpumalanga

The northern section of the Saddleback Mountains has escaped afforestation with pines. It harbours some significant populations of *Aloe chortolirioides* representative of the distribution of the plants across the mountain slopes prior to afforestation. The plants are most common on east-, north- and west-facing mountaintops. They tend to be concentrated mostly on the upper third of the highest slopes, just below the summits on the hottest slopes facing north and west.

The aloes come into flower in years when the grassland is burnt, with the peak of the flowering season between June and August. Fires burn erratically on these mountainsides and some slopes escape burning altogether. The fires burn back upon themselves when the wind changes direction and, unless the fire is extensive, the flames do not approach an unburnt area again during the course of the same fire.

A good example of the biological processes at work at this locality was observed during the winter and spring of 2004. In June, a fire had burnt an extensive area of the habitat. An east-facing slope with a population of some 50 plants had escaped burning except for the bottom section of the slope owing to changes in the direction of the wind during the course of the fire. On 24 August, the plants on the warm eastern aspect that had been burnt were in the final stages of liberating seeds. Most plants that had escaped burning were fully dormant. Plants growing on the north-, west- and south-facing slopes were all about two weeks away from liberating seeds. These plants, all growing in windy, colder positions had flowered slightly later than those growing in places lower down on the eastern aspects, which are warmer at this time of the year and less exposed to wind.

The flowers on the stems of individual plants had been well pollinated over an extensive area of a few square kilometres. The countryside was clear and open and the orange-red flowers growing in scattered clumps were attractive to the full range of avian and insect pollinators. The flowering aloes were particularly conspicuous against the background of blackened veld after the grass fire.

Some 73 stems with ripening seeds were examined randomly across the population of plants. They were close to liberating seeds and at this time it is possible to examine levels of parasitism. The seed on 11 stems was in the process of being eaten by beetle larvae. About 20% of seed capsules had the characteristic small round holes indicating parasitism by seed chalcids.

This fire followed by a good flowering season and seed set with limited parasitism is a rare event. Gusting windy days in the spring ensured furthermore that much of the seed was widely dispersed by wind action.

Colonies in the region of extensive afforestation between Barberton and the Swaziland border

The plants growing in areas heavily afforested with timber trees, mostly pines, are gradually declining in numbers as the years go by. Fires are infrequent, with some plants escaping burning for up to a decade. Other plants are regularly burnt if they grow on firebreaks adjacent to plantations. The firebreak areas are burnt during the dry winter. This stimulates the plants to flower and some seed is set when there are pollinators in the area. The seed that germinates produces seedlings, which get burnt the following winter

Right: ***Aloe chortolirioides*** amongst rocks on the mountains between Barberton in Mpumalanga and the Swaziland border, in full flower after a winter grass fire.

at a time when they are not yet large enough to resist fire. Few, if any, seedlings are found on most firebreaks unless wedged between rocks where they escape the worst effects of the fire. Populations on the firebreaks consist mostly of old plants that were present in the grassland at the time the area was afforested.

The groups of aloes that exist in the plantation areas are not burnt in any synchronised manner as they are when fires occur in the unforested area closer to Barberton. This isolates groups of flowering aloes from one another. In former times they grew in a huge area of grassland subject to extensive simultaneous burning. Irregular fires in isolated grassland have the effect of restricting potential flowering seasons and limiting opportunities for the recruitment of seedlings to the different populations. The plants are furthermore probably not be visited by many pollinators, resulting in a negligible seed set.

A colony of *Aloe chortolirioides* growing on a hill surrounded by alien tree plantations was visited from noon until midafternoon on 24 August 2004. The grassland had been burnt earlier in the winter and the plants were in flower. Plants on the sunny slopes had finished flowering while those on the southern aspects were just starting to flower. Only one pollinating insect was observed from 12:00 to 15:00, a time when pollinators are usually at their most active. This insect, a carpenter bee (*Xylocopa caffra*), visited one plant with about 20 flowering stems growing in full sunlight. The aloes that had finished flowering nearly all lacked characteristic swellings at the base of the flowers. When present, these swellings indicate the beginning of seed formation.

Growth cycle and the role of fire

Aloe chortolirioides is a fire-dependent species for flowering, seed distribution and seedling development. Fire is not only essential for the reproduction of this grass aloe, but also for maintaining adult plants in a healthy state. It burns away dead branches on the older plants as well as loose stem material prone to fungal attack.

It is likely that the species is very long-lived and that conditions conducive to abundant flowering and a good seed set followed by plentiful rain to promote germination, are rare. The colony on the Saddleback Mountains has experienced these conditions twice in the 20-year period during which the plants were studied.

In the winter months *Aloe chortolirioides* is indistinguishable from the short, dry grass amongst which it grows. The dry aloe leaves are the same size and shape as dead grass blades. They are also often borne at the same height as the tips of the grass blades. This flammable material ensures that the tops of the aloe stems from which the flowers emerge, are thoroughly burnt during grass fires. *A. chortolirioides* produces flowers within a few weeks after a grass fire. New leaves start to emerge at the same time and continue to grow until the early summer when a point of maximum leaf production is reached. Aloes in populations that have escaped fire remain dormant until late into spring. They generally start producing some new leaf growth at the time the main summer rains begin, from early October to early November.

The aloes start to enter dormancy in May, at a time when the grassland is beginning to dry out and when day length decreases and temperatures drop.

Seedling plants are recruited to the populations in any numbers only after grass fires, but in a dry summer many succumb after germinating. Young plants of approximately the same age are found in some populations where flowers are still visited by pollinators. These plants indicate years when there were climax conditions for the regeneration of the species.

Status of wild populations

There are a few remaining populations of this species where the biological processes affecting their life cycle have not been significantly altered. Plants growing in heavily afforested areas near Barberton are mostly old ones, with very limited possibilities of regeneration from seed. It would be useful if attempts could be made to conserve populations where fire cycles are still a regular part of the grassland ecology.

Assessments of the status of this species have to take into account not only the total numbers of plants in given populations, but also the integrity of the biological processes that affect their life cycles. Most colonies in timber-producing areas are not viable in the long term without significant and probably costly management programmes.

Cultivation, propagation and horticultural potential

Aloe chortolirioides is an elegant species suited to growing in containers or on sunny rockeries in areas with summer rainfall. The plants do well in terracotta pots, particularly those with a rough finish that absorbs excess moisture after watering.

An ideal cultivation mixture consists of half sieved, sandy soil and half fine river sand to which some very well-rotted veld grass compost has been added. It is often a good idea to place rocks around the base of the stem to conserve surface moisture. These conditions approximate those found in the habitat of the plants.

The plants rarely flower in cultivation unless burnt towards the end of winter or in early spring. Dry grass from the veld should be heaped over the stems, particularly around the dead leaves from the previous growing season. This practice ensures that the leaf-producing parts of the stems are thoroughly burnt. Damage to the container in which the plant is grown can be avoided by cutting a hole the same size as the top of the pot in a piece of stiff board or asbestos. The dry leaf material will then burn only above the board without discolouring or overheating the pot. *Aloe chortolirioides* should not be burnt every year, rather every second or third season. Too frequent burning may destroy the new shoots the plant has grown the previous summer.

Flowers have to be hand-pollinated to ensure that some seed is formed. Seeds should be harvested as soon as they ripen and sown in late spring or early summer. They should be lightly and evenly scattered across the surface of the propagation trays. The same mixture can be used for propagation as for the cultivation of adult plants. Seed trays exposed to rain yield the best germination results. The problem with this practice is that heavy rain and hailstorms destroy germinating seeds and recently germinated seedlings. This difficulty can be solved by covering the trays with fly screen netting. It keeps off the hail and minimises the effect of heavy raindrops. Not all *Aloe chortolirioides* seed germinates the first season and the percentage germination after the initial sowing varies greatly from one batch of seed to the next.

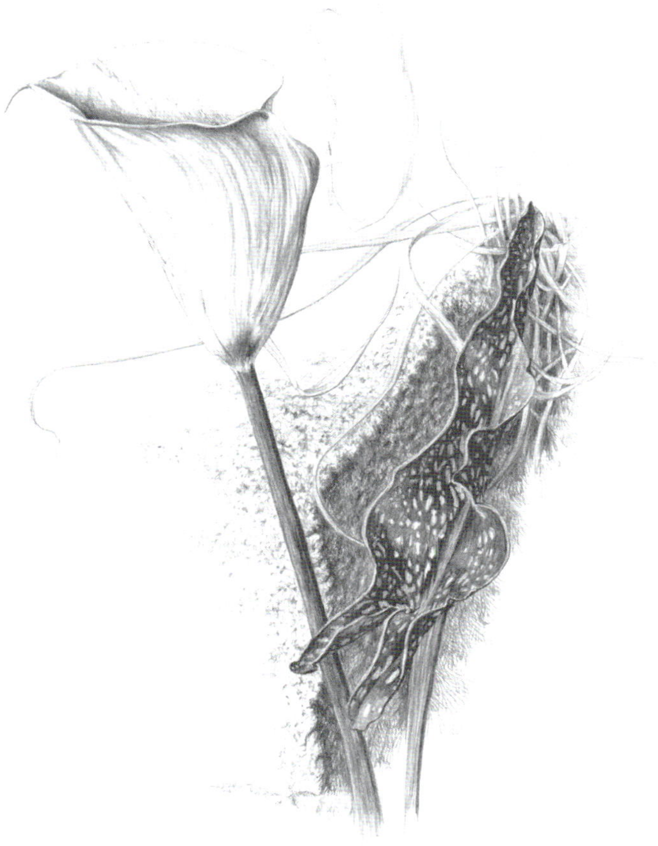

Zantedeschia jucunda is frequent along the summit of the central and southern summit slopes of the Leolo Mountains in Sekhukhuneland in Limpopo Province. The plants even grow in vertical rock cracks with minimal soil. *Aloe cooperi* often grows close to *Z. jucunda* but cannot utilise the rock crevice habitat, preferring deeper soil amongst boulders.

Adult plants and trays of seedlings do best in morning sunlight or light, dappled shade where conditions of daylong sunlight are provided. Mature plants and seedlings require plenty of free air circulation to remain healthy. Seedlings should be burnt in the same manner as the adult plants once they reach a height of about 2.5 cm. Burning keeps the growth habit compact and helps to eliminate fungal spores from the stems. It also encourages the young plants to branch.

Aloe chortolirioides has much horticultural potential, particularly for ornamental container planting, on account of its branched habit and tufts of grass-like leaves. It is, unfortunately, not the easiest of species to cultivate successfully and seed is usually in short supply. This is one of the grass aloe species that would merit production by tissue culture techniques, should they prove possible for this species.

4. *Aloe cooperi*

Aloe cooperi Baker

Aloe cooperi is one of the widespread grass aloes found in many parts of KwaZulu-Natal, Mpumalanga and also in the southwestern section of Limpopo Province. The species is variable, as can be expected from a grass aloe found in different habitats over a wide geographical area.

One of the problems facing taxonomists is whether or not this variation should be formally recognised. The species can either be treated as geographically variable or recognised in terms of subspecies or variety. At the present stage of knowledge development, the situation remains to be clarified. Glen & Hardy (1987: Plate 1944) describe *Aloe cooperi* subsp. *pulchra* based on minor floral and foliage differences from *A. cooperi* subsp. *cooperi* as well as preferences for different habitats in localities where the two occur sympatrically. The flowering time in subsp. *pulchra* is furthermore much later, in April and May, at a time when plants of typical *A. cooperi* have long since finished flowering and already liberated their seeds.

The problems arise, however, with the fact that most variation in *Aloe cooperi* is not clear-cut. This is discussed on the basis of some examples drawn from field data collected for this book. One of the main characteristics that are said to separate *A. cooperi* from other large grass aloes are the distinctively keeled leaves which appear V-shaped in cross section. A large population of these plants near Amsterdam in Mpumalanga has plants with strongly keeled leaves amongst those with rounded or flatter leaves. They may be weakly keeled or lack keels altogether. These plants also flower about a month earlier than most *A. cooperi*, with a peak flowering period in the first half of November. The floral structure even varies, with racemes either conical or nearly capitate.

Aloe cooperi growing on the southern section of the Leolo Mountains is another example of significant variation. Plants on the summit of the mountain are often only 15–20 cm tall and grow in rocky places, often in the vicinity of seepage areas. Plants lower down grow on boulder outcrops in pockets of deep soil. Some of these are very robust, with flowering stems nearly 1 m tall. These aloes are as the crow flies less than 1 km away from the gracile specimens on the summit of the mountain. The two examples of *A. cooperi* on the Leolo Mountains have the same flowering season.

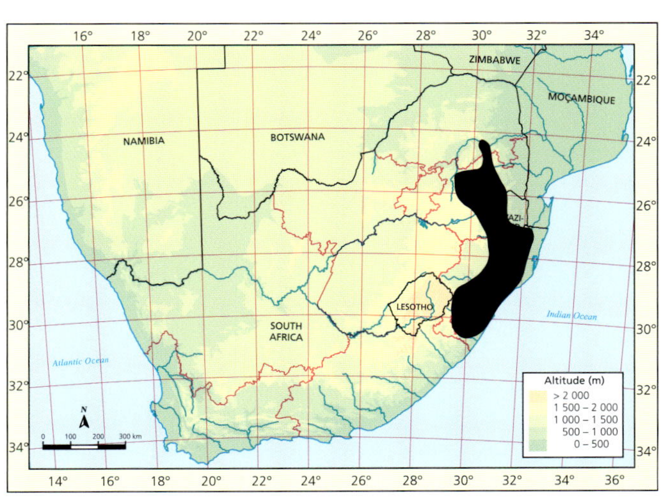

Distribution of *Aloe cooperi*

The formal recognition of variation in *Aloe cooperi* is likely to be made difficult by the paucity of clear-cut boundaries normally used to describe subspecies and varieties. It is necessary that the *A. cooperi* complex is studied through detailed sampling. This is likely to lead to well informed decisions as to whether or not it is feasible to describe variation within the concepts of subspecies and varieties.

A decision was made not to recognise subspecies in *Aloe cooperi* at the present time. Once the concept of *A. cooperi* has been theorised with reference to detailed fieldwork, the status of *A. cooperi* subsp. *pulchra* could be confirmed. Decisions of this nature may be made as a result of recognising further variety in *A. cooperi*. Alternatively, the existing subspecific rank may be confirmed as the only variation in the *A. cooperi* complex with sufficiently constant characters to merit formal recognition.

Aloe cooperi has been studied in several parts of its distribution range—the Leolo Mountains in Sekhukhuneland, the hills in the Chieftain's Plain, Roossenekal and Tonteldoos areas, the Amsterdam area and the vicinity of the Bivane Dam, consisting of the hilly country southwest of the confluence of the Pongolo and Bivane Rivers. The study areas are fully representative of the modern environments in which this grass aloe finds itself.

Near Amsterdam in Mpumalanga

Amsterdam is a small town in southeastern Mpumalanga near the border between South Africa and Swaziland. Much of the countryside is heavily afforested with alien timber trees except for some large areas with rocky hills not suitable for the establishment of plantations. Some of these habitats are rich in bulbs, succulents and other grassland flora. Most of them are, however, in various stages of transformation by invasive black wattles (*Acacia mearnsii*) and to a lesser extent by silver wattles (*A. dealbata*).

A river valley surrounded by steeply sloping hills has some extensive populations on the shady south- and southwest-facing slopes, while the hot, sunny north- and northwest-facing slopes support one of the largest remaining colonies of *Aloe integra* in South Africa. Black wattles have invaded mainly the southern hill slopes as well as the river valley and its tributaries. The hotter slopes have infestations of kikuyu grass and various alien weeds such as *Verbena aristigera* in areas that have been over-grazed.

Aloe cooperi is found mostly on the upper third of hill slopes, often occurring in large numbers amongst thick grass and between rocks. The plants are less frequently found in the river valley. In this situation they become very large and robust owing to the deep soil which permits maximum root development.

The aloes flower very well the first season after a winter grass fire. The flowering performance of the plants declines sharply when the veld has not been burnt for several years in succession. Mass flowering usually leads to good seeding, with large amounts of seed liberated into the environment when germination conditions are also at their optimum.

The species declines in numbers where it is surrounded by invasive wattles. The grass fires lead to a rapid expansion of these wattles since their seeds germinate in large numbers after burning. Younger aloes are destroyed by the intense heat produced when stands of wattles get burnt during the course of the fires.

On the Leolo Mountains in Limpopo Province

Aloe cooperi is usually a gracile dwarf plant where it occurs sparsely amongst rocks on the summit of the Leolo Mountains, often together with the beautiful yellow arum (*Zantedeschia jucunda*), a Leolo Mountain endemic. The aloes grow next to rocks where they escape trampling by livestock. They also occur in rock fissures, growing amongst short tufts of grass on the vast norite rock domes that characterise the summit of the Leolo Mountains.

On the lower slopes of the mountain the plants are very large and robust, growing mostly between boulders. They are confined to this habitat as the mountain slopes are heavily grazed and the aloe stems are readily damaged by the hooves of livestock. The difference in size between plants growing on the summit and those on the lower slopes seems to be caused by little more than different growing conditions—the soil is shallow on the mountain summit where *Aloe cooperi* is found, but deep on the lower slopes.

In the Chieftain's Plain, Roossenekal and Tonteldoos areas of Mpumalanga

Aloe cooperi is widespread but not common in the Chieftain's Plain, Roossenekal and Tonteldoos areas. The plants are subject to all the modern pressures in South Africa's grasslands. Near Roossenekal the aloes have been destroyed by recent mining operations. They are subject to trampling by grazing livestock throughout the area, but particularly on the Chieftain's Plain, and their habitat is being transformed in some parts by black wattles.

Near Roossenekal the aloes have adapted to a peculiar habitat, namely leaf compost spread over the tops of large boulders. The compost is replenished annually from the deciduous shrubs that form a dense canopy above these boulders. The physical differences found in these plants, namely long stems that spread over the rocks and flaccid, prostrate leaves, are caused by the shallow compost and shaded conditions. Aloes growing only a few meters away in patches of grass, exposed to direct sunlight, conform in all respects to the typical *Aloe cooperi*.

In the Chieftain's Plain area, *Aloe cooperi* is confined mostly to boulder outcrops owing to the fact that the habitat is heavily grazed. In this respect the distribution of the plants at this locality is little different from that on the Leolo Mountains about 60 km to the northwest.

Around the Bivane Dam near the confluence of the Bivane and Pongolo Rivers

The Bivane Dam is situated to the east of Vryheid in KwaZulu-Natal. The area consists of large hills with deeply incised river valleys. *Aloe cooperi* grows sparsely on steep, grassy slopes, in thin soil amongst rocks. The plants are usually about 20–30 cm tall in habitat, but become very large when grown in a garden and planted in deep soil.

Right: ***Aloe cooperi*** growing in a grassy crevice between boulders on the Chieftain's Plain east of Roossenekal in Mpumalanga. The plants come into flower when the grasses are in active growth, the flowers usually well elevated above the grass blades where they are conspicuous to pollinators.

The aloes are confined mostly to the vicinity of seepage areas on very steep hillsides since the remainder of the habitat is too arid for their survival. They grow in an area of less rainfall than most localities where *Aloe cooperi* is found. Dense mists that occur in summer do, however, augment the moisture source available to the plants.

Growth cycle and the role of fire

Aloe cooperi is a summer-growing species sprouting leaves with the beginning of the main summer rains in November. Leaf growth is rapid once it has been initiated. Most plants flower in December and January and their seeds ripen in late January and February. The aloes enter dormancy in April and May and at this time the leaves wither and dry out. Late rains and mild weather in the early winter may delay the onset of dormancy, the leaves then remaining green until late May or early June.

The plants usually grow amongst tall grasses. The cover becomes very dense a few years after the last fire. At these times, flowering performance tapers off until the next fire. Winter grass fires always stimulate mass flowering the following summer. Fires are essential for creating conditions conducive to the germination of seeds of *Aloe cooperi*. The grass cover after a fire is much less dense than in the inter-fire years, providing the right amount of filtered sunlight needed for successful development of seedlings.

Status of wild populations

The species is widespread but is rarely found in large populations. Much of its former habitat in KwaZulu-Natal has been destroyed by timber plantations. Transformation of its grassland habitat by invasive alien vegetation is becoming a problem in parts of KwaZulu-Natal and Mpumalanga.

The status of *Aloe cooperi* will have to be assessed again if formal recognition is afforded to the variation found in this species. One of the main problems involved with such a reassessment is the fragmented habitat occupied by the aloes. This is likely to affect the authenticity of any divisions made within the currently accepted species concept of *A. cooperi*. In this respect the circumstances are similar to the problems plant taxonomists will face with respect to several South African grass aloe species.

Cultivation, propagation and horticultural potential

Aloe cooperi is widely used in South Africa as a garden subject and for more extensive landscape features. The plants are readily propagated from seed and grow fast, reaching flowering size more quickly than other grass aloes of comparable size. The aloes can be grown in large, deep containers or in garden beds. They should be planted in well-drained situations in gardens, preferably on sloping ground or in a rockery.

Seeds should be sown in a well-drained, porous mixture. A good sowing medium consists of half sieved sandy soil to which equal parts of silt or river sand and finely sieved well-rotted compost have been added. Seeds are best sown in large, deep seed trays or pots which permit the aloes to develop their deep root systems. Seeds should be lightly and evenly scattered across the surface of the soil and kept moist during germination and for the remainder of the summer while the young plants are developing. Watering should be gradually reduced during late summer and early autumn. The seedlings should be kept mostly dry in winter, with an occasional watering during mild weather. Regular watering should be resumed in spring.

Aloe cooperi can also be grown from cuttings which should be taken in spring or autumn and allowed to dry out for a few days. They should be struck in coarse, gravelly sand.

This species has a great deal of horticultural potential, some of which has been realised in South African horticulture. Selections can be made in favour of plants with the most copiously white-spotted leaves and the largest flowers. The less robust examples of *Aloe cooperi* are well suited to ornamental container planting.

Right: ***Aloe cooperi*** on the rocky hills near Amsterdam in Mpumalanga in midwinter 2004 when there was a huge mass of dead grass in the veld, ready for a hot veld fire. Such a burn occurred in late winter, resulting in mass flowering on the slopes the following summer.

5. *Aloe craibii*

Aloe craibii Gideon F.Sm.

Aloe craibii was only recently described by Smith (2003). The plants occur in convoluted mist belt country in the mountains to the southeast of Barberton in Mpumalanga. This area falls within the Barberton Centre of Floristic Endemism (Van Wyk & Smith 2001: 116–119). The plants have so far been recorded from very few localities, all of which lie within or on the boundaries of areas extensively afforested with eucalypt and pine plantations. The species has not been recorded from elsewhere but may in due course be found on the highest peaks in the Songimvelo Nature Reserve and the nearby Malolotja Nature Reserve situated along the border area between Mpumalanga and southwestern Swaziland.

Smith considered *Aloe craibii* to be most closely related to *A. integra*. It also, however, has some similarities to *A. linearifolia*, a species found in coastal grassland in southern KwaZulu-Natal. *A. integra* and *A. linearifolia* sometimes, but not always, grow in mist belt areas while present indications are that *A. craibii* is a mist belt endemic.

The species occurs mostly at altitudes ranging from approximately 1 500–1 800 m. The plants frequent short, rocky grassland occurring on both serpentinite and sandstone substrates. The aloes have long, thick, fusiform roots and this restricts them to large pockets of soil amongst rocks, steep grassy slopes with deep soil and, less frequently, localities on rocky outcrops where the roots are able to spread.

Grass aloes that frequent mist belts have either large, fusiform roots or thin, fibrous root systems. Those with fibrous roots are able to colonise rocky areas as well as cliff faces and can utilise habitat unavailable to grass aloes with deep root systems. Species with deep roots tend to be sparsely scattered over suitable mist belt habitat, for example *Aloe chortolirioides*, *A. craibii* and *A. woolliana*. Species such as *A. nubigena*, *A. soutpansbergensis* and *A. thompsoniae* have thin, spreading root systems, allowing them to become abundant in most of the mountaintop habitat where soils are thin.

Aloe craibii has been studied at two sites in mountainous mist belt grassland southeast of Barberton—the type locality and the upper slopes of a large mountain close to the border between Mpumalanga and Swaziland.

In the type locality

The type locality consists of a large, rocky hill surrounded by pine plantations on most sides. The north-facing slopes are arid, with thin soils, and home to various xerophytic plants such as *Euphorbia pulvinata*. The east-, west-, south- and southwest-facing slopes contain a large amount of rocky grassland as well as seepage areas with dense grass cover and deep soil. The eastern side of the hill has little grassland. Much of this habitat consists of broken, rocky cliffs. *Aloe craibii* currently occurs on the south-, southwest- and west-facing slopes. The plants were once scattered across the northwestern slopes but in recent times they have been afforested with pines. The aloes growing in rocky areas have short stems, often single and more rarely with one or two branches. Those growing in deep soil have five to eight branches, most of which produce flowering stems.

The habitat is currently being degraded by annual and perennial alien weeds often associated with soil disturbance at the edge of timber plantations. They are beginning to invade the grassland, particularly after fires have periodically cleared the habitat. Another problem is invasion of the habitat by pine trees (*Pinus patula*) that have sprouted from seeds blown into the grassland from surrounding timber plantations.

The recruitment of young plants to the population has always been slow owing to the irregular burning of the grassland. The type locality is in a fire prevention zone, the only permitted burning consisting of fire breaks on the perimeters of timber plantations. The lack of fire has had the effect of restricting seedlings to the most rocky habitat where the grass cover is at its thinnest. The most suitable habitat is covered with dense grass for several successive seasons in the years between the infrequent fires.

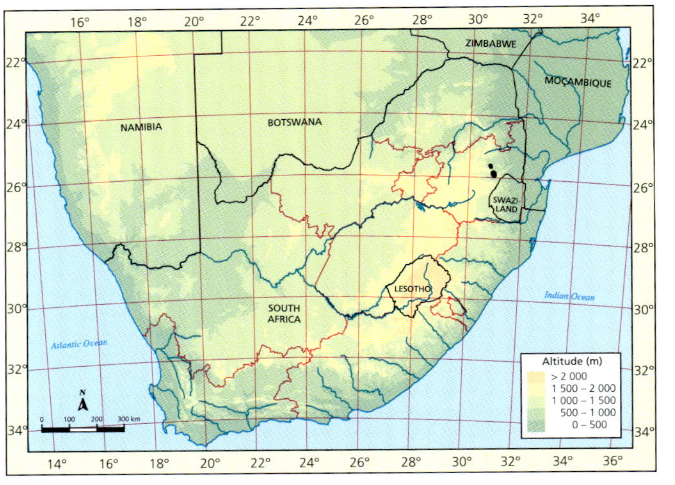

Distribution of *Aloe craibii*

Near the border between Mpumalanga and Swaziland

Aloe craibii has been studied for about 10 years near the border between Mpumalanga and Swaziland. This area is close to the eastern boundary of the Songimvelo Nature Reserve and just outside the southwestern border of the plantation belt. Winter grass fires are more regular at this locality but the grassy habitat suitable for the aloes is restricted mostly to the west- and southwest-facing slopes. The northern, northwestern and northeastern slopes contain mostly cliffs and rocky outcrops with thin soil. This situation is typical of much of the remaining habitat between Barberton and Swaziland. Many areas of mountainous grassland suitable for this species are now covered with timber plantations.

Aloe craibii occurs on the central and upper southwest-facing slopes with a good deal of deep soil with short grass cover. This is ideal habitat for the species and most of the plants are multistemmed, a growth habit indicative of good soil depth. The aloes are thinly but evenly scattered throughout the habitat.

The grassland is lightly grazed throughout the year, mainly by cattle and horses, and is subject to regular

Right: ***Aloe craibii*** in flower at the type locality in Mpumalanga between Barberton and the Swaziland border, shortly after the main flowering period in early April. The exclusion of fire and pollinators from its habitat is likely to lead to the extinction of this species during the course of this century.

G Condy
2003

burning. Fires usually occur when the grass is long after good rains the previous season. Recruitment of seedlings to the population is slow as a result of an erratic and unpredictable seed set. The habitat occupied by this species is often subject to prolonged dense misty weather at flowering time. Few, if any, pollinators visit the flowers under these conditions.

The base of the large hill where most of the *Aloe craibii* plants grow is being seriously invaded by black wattles (*Acacia mearnsii*). They are progressively spreading along seasonal seepage areas and can in due course be expected to transform much of the grassland at this locality.

Growth cycle and the role of fire

Aloe craibii is one of the latest species to start its growth cycle. The leaves start to emerge after the first regular successive summer rains, at any time between the beginning of October and early December. Leaf growth continues until February. Flower buds start to form from late February and the plants come into flower in mid-March, extending to early April. Seeds are liberated into the environment late in autumn or at the beginning of winter, mostly during the second half of April and the first half of May. The flowering period is a few weeks earlier in seasons when abundant summer mists and rainfall follow a grass fire the previous winter. Seeds are dispersed mostly by wind and rain, their first chance to germinate being the start of the following rainy season at the beginning of the next summer.

The species is dependent on fire cycles for its health and regeneration. Fire clears the habitat from dead grasses and dry leaves of bracken (*Pteridium aquilinum*). It also redistributes nutrients in the mountain grassland and burns dead leaves off the aloe stems. Accumulations of dead leaves harbour pests and fungal spores and after fires the blackened externally charred stems act as a fungal repellent. Fire clears the most favoured habitat, namely areas of deep soil with thick grass, from clutters of moribund vegetation. These conditions promote flowering as well as dispersal and germination of seeds.

The flowers are pollinated by black sunbirds (*Nectarinia amethystine*) and malachite sunbirds (*Nectarinia famosa*). Insects also pollinate the flowers but they are now irregular visitors to areas extensively afforested with timber. This problem is exacerbated by prolonged misty weather which is frequent at flowering time. Pollinators are unable to visit the flowers during heavy mists.

Status of wild populations

Aloe craibii is a rare plant as much of its former habitat has been destroyed by afforestation. The remaining areas where the plants grow are in the process of being degraded and transformed by weeds and invasive alien vegetation.

A survey is required to assess the status of plants and also to decide which areas are best suited to the long-term preservation of the species. The future of this aloe will be more secure if good populations are found in the Malolotja Nature Reserve in Swaziland and the nearby Songimvelo Nature Reserve in Mpumalanga.

Habitats where the species occurs in plantation areas will have to be artificially managed to secure the long-term survival of these aloes. This will involve periodic burning of the grassland and removal of weeds and alien vegetation. The population dynamics of the different colonies of plants are likely to need monitoring to ascertain whether or not some pollination of the flowers occurs. It may in due course prove desirable to translocate some plants to the highest peaks of the mountains in the Songimvelo Nature Reserve, which is relatively close to the type locality and virtually on the boundary of the colony close to the Swaziland border. The long-term survival of the plants is precarious unless they occur naturally in one or both of the two nature reserves.

Cultivation, propagation and horticultural potential

Aloe craibii is scarce in cultivation and observations on its performance under cultivated conditions are rather limited. Plants should be grown in large containers, preferably clay pots. A good amount of drainage material such as coarse stones or fragments of broken clay pots should be packed at the bottom of the containers. Above this there should be a thin layer of small stones or coarse gravel. A good mixture in which to cultivate the plants consists of one third coarse sand, one third gravel and one third finely sieved peat moss or very well-rotted, finely sieved, acid leaf mould.

The aloes should be well watered from late spring until late autumn or early winter, October to about early May in the southern hemisphere. They should be kept mostly dry for the duration of the winter dormancy period, with an occasional light watering during mild weather.

The species also grows well in garden beds exposed to about half a day's sunlight. The situations in which the plants are cultivated have to be well drained. These requirements are best met by digging a garden bed in a sloping area with sandy soil. The root run has to remain undisturbed in friable soil. This is facilitated by placing large stones and coarse organic mulch around the area where the plants are growing.

The aloes should be burnt once every two to three years towards the end of the winter by heaping dried veld grass around their stems and igniting it. This practice keeps the plants healthy and promotes flowering towards the end of the following growing season.

Aloe craibii is easily propagated from seed. Seeds should be sown in the same mixture used for cultivating mature plants in containers. They should be lightly and evenly scattered across the surface of the soil and just covered with sand or silt. Large, deep containers are the most suitable since it allows the seedlings to develop a deep root system unfettered by shallow soil in conventional propagation trays. Seeds should be sown in spring or early summer. The containers should be well watered but not kept too damp. Regularly spaced watering should continue while the seeds are germinating and afterwards until autumn or early winter when the leaves start to die back, as temperatures drop and the seedlings enter dormancy.

6. *Aloe dominella*

Aloe dominella Reynolds

This multistemmed species with tufts of grass-like leaves bears a rather close resemblance to *Aloe chortolirioides* with respect to its habitat preferences, growth form and the fact that burning of the veld is required to stimulate mass flowering.

Aloe dominella is found mostly in short grassland in several districts of central KwaZulu-Natal, particularly the Bergville, Dundee, Estcourt, Mooi River and Weenen areas. The habitat frequented by the plants is hilly or gently undulating, in the region of about 1 500 m. This is the only species amongst the three multistemmed South African grass aloes that does not occur in habitats subject to prolonged and frequent mists in summer.

The grassland in which *Aloe dominella* occurs is mostly cattle ranching country. Some of the country in which the plants are found is situated within communal grazing areas. The grassland is heavily grazed by cattle, sheep and goats in most of these areas. *A. dominella* is rarely common in any areas regularly grazed by domestic stock. The multistemmed plants are easily damaged by the hooves of grazing animals. Some of the habitats are in the initial stages of invasion by black wattles (*Acacia mearnsii*), which have transformed much of the grassland habitat in parts of central KwaZulu-Natal.

The species is partial to steep, dry, rocky slopes as well as areas of short, dense, undulating grassland. In most of the areas where the species is found it is well represented in fenced road reserves but scarce in the surrounding grassland, owing to grazing pressures. In one area southeast of Estcourt in KwaZulu-Natal, *Aloe dominella* inhabits very steep, grassy slopes where grazing cattle have limited access and the plants here have become large owing to the lack of disturbance by grazing animals.

There are currently a number of land claims in KwaZulu-Natal and the method of land usage in most of these areas is set to change from large cattle ranches to small-scale pastoralism. Livestock numbers would increase and the grasslands would become more degraded owing to increased grazing pressures. This can be expected to lead to a further decline in numbers of the species.

Aloe dominella has been studied at three localities in KwaZulu-Natal, namely southeast of Estcourt, southeast of Dundee near Rorke's Drift, and close to Bergville in the eastern foothills of the Drakensberg. These localities represent the range of habitats occupied by this grass aloe.

Southeast of Estcourt

The plants grow on steep and more gentle north- and northwest-facing hillsides south of Estcourt in cattle ranching country. They are also found in the reserves of roads that run adjacent to the farmland. This provides a good opportunity to study the numbers of plants in the road reserve, which is not grazed by livestock, and the surrounding grassland areas.

The aloes are best represented on the steep slopes where they are difficult for grazing cattle to reach. At the time the data were collected, there were no goats grazing these hills. Should goats have been present, the plants would have been damaged and trampled as they are elsewhere in places where mixed livestock is kept.

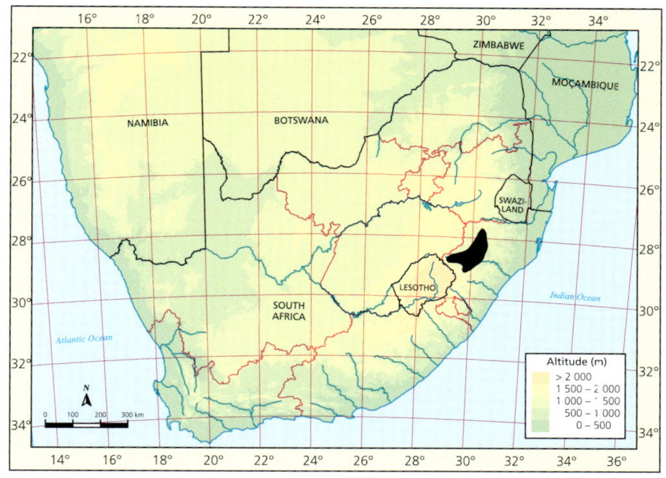

Distribution of *Aloe dominella*

The clumps grow close to one another on the steep hillside, usually about 1 m apart. They are most frequent on rocky ground or beside sheets of exposed sandstone, near a seepage area. About 700 m further away, the hills are much less steep and the road reserve is 20 m or more wide in places. In these situations the reserve supports large and small clumps of *Aloe dominella*. The adjacent farmland, over the fence, has far fewer plants and then mostly wedged in between or beside large rocks. In the winter months the aloes are characterised by extremely thick mats of prostrate, highly flammable dead leaves which are thoroughly burnt during the course of grass fires.

There are few seedling plants, even in the road reserve. This is related to the fact that this species is not a prolific seeder and young plants are readily destroyed if the grassland is subject to fire in consecutive seasons.

In the Bergville area

Near Bergville, *Aloe dominella* occurs in grassland where mixed stock farming is practised on communal grazing lands. Goats form a large component of the herds and they regularly graze the rocky hills where the plants occur. The habitat is characterised by large sheets of exposed sandstone and boulder outcrops. The aloes are absent from rocky ground around the rock sheets, a common habitat for this species elsewhere. They occur in very limited numbers amongst large rocks where it is difficult for goats to trample or graze them.

Aloe dominella has virtually no long-term prospects for regeneration in this type of habitat. The heavy grazing has, however, benefited other types of plants such as *Pelargonium bowkeri*, which grows in the same habitat. The pelargoniums frequent areas where the grass cover is always very short. This confines them to rocky outcrops and patches of soil on small cliffs. Regular grazing keeps the grass short all year round and permits the plants to spread into most suitable places in the habitat.

Near Rorke's Drift

Aloe dominella occurs in rocky hills to the west of Rorke's Drift. Much of this area consists of undulating hills. The slopes are gentle and easily accessible to grazing livestock. The species favours rocky road reserves in this area, which are fenced off from the surrounding farmland and periodically burnt during the course of veld fires. The accompanying illustration was painted in such a road reserve where the plant grew alongside several other clumps of these aloes. A thorough search was carried out in the adjacent farmland, but no plants of *A. dominella* were found. The grassland was well grazed by cattle at the time the research data were collected, explaining their absence from the surrounding farmland.

Growth cycle and the role of fire

Aloe dominella is a summer-growing and winter-dormant species. The plants respond directly to winter grass fires and flower mostly towards the end of winter and in early spring, between late July and late September.

The grass cover where the plants grow frequently becomes taller than their highest stems. In addition to the dense surrounding grass cover, the aloes accumulate a great deal of dead leaves when they have not been burn for a few years in succession. These conditions ensure that the stems in the aloe clumps are very thoroughly burnt during grass fires. The stems are distinctively black after burning. The plants harbour many pests during inter-fire years, including mealybug and woolly aphids. Fires rid the plants of these infestations and are very necessary for maintaining the health of the aloes.

Aloe dominella is regularly visited by pollinators for the duration of its flowering period, which lasts a few weeks. The seed set, however, is rather variable. The time of the year when the plants flower is often very windy in the grasslands of central KwaZulu-Natal, sometimes reducing opportunities for insects to visit the flowers for a few days in succession. *A. dominella* flowers poorly during the inter-fire years and plants growing in long grass may fail to flower at all.

The seed is dispersed between September and October. The aloes restrict mass flowering to years when they are thoroughly burnt, ensuring that any seed set after flowering has the optimum conditions for dispersal and germination. The thinner grass cover after fires furthermore shades the developing seedlings without blocking out too much sunlight.

The optimum conditions for regeneration of this species are fire cycles once every three to four years. Fires are, however, random and unpredictable. *Aloe dominella* seedlings are confined to the vicinity of rocky outcrops in areas where grass fires occur during most winters.

Status of wild populations

Aloe dominella is widespread in the grassland of central KwaZulu-Natal, but rarely plentiful in any locality. Its numbers have been reduced by too frequent burning and trampling by livestock. Road reserves play a significant role in protecting this aloe. The plants in these reserves also serve as a source of seeds, the vehicle through which this species can re-establish itself in surrounding farmland. Its numbers may decline sharply if land use patterns change to mixed herds on communal grazing lands.

Some *Aloe dominella* habitat southeast of Estcourt has been transformed by black wattle infestations. These wattles are a problem in much of the grassland in the Estcourt and Mooi River areas of KwaZulu-Natal. The distribution of *A. dominella* lies mostly outside the region used for timber production. This situation serves to reduce the variety of environmental pressures these plants face in the long term.

Cultivation, propagation and horticultural potential

Aloe dominella is relatively easy to grow in cultivation, but difficult to flower. The plants should be planted in large pots, in well-drained, sandy soil. The large containers afford the plants the opportunity to develop their extensive root run. A good cultivation medium consists of one third coarse gravelly sand, one third sieved, sandy soil or sandy loam and one third sieved, well-rotted compost.

The plants thrive in positions of day-long direct sunlight or lightly dappled shade. They require plenty of free air circulation to keep them healthy and free of pests. This species is ideally suited to pockets of deep soil in a rockery or stretches of natural well-drained, rocky grassland in a veld garden that also have some areas of deep soil.

The aloes should be thoroughly watered at intervals during the summer months and kept dry in winter. Watering may commence shortly before the end of winter, in July and August in summer-rainfall areas of South Africa.

The species rarely flowers in cultivation unless it is periodically burnt amongst a small pile of dead veld grass in late July, towards the end of winter. The grass should be heaped around the aloes and ignited, ensuring that there is sufficient flammable material to burn the stems and the dry foliage. Plants burnt in late July can be expected to come into flower during the second half of August.

Aloe dominella is easily propagated from stem cuttings. Cuttings should be broken off the parent plant toward the end of winter. They should be left in a dry, shaded place to heal for a few days and then placed in moist sand. Roots usually form within a few weeks and the cuttings are ready for transplanting at this stage.

The species can also be grown from seeds. It sets little seed in cultivation without the aid of hand pollination. This aloe also requires cross-pollination to produce a good seed set. Seeds should be lightly and evenly scattered across the surface of the soil and just covered. Large pots or deep propagation trays are best for sowing seeds. The same mixture can be used for sowing seeds as for cultivating adult plants. Containers used for propagation purposes have to be well drained. Seeds should be sown in spring and well watered periodically, after they have been sown and throughout the germination and seedling development period. The germination mixture should be damp, but not soaked since too much moisture precipitates damping off.

The young plants usually start to show signs of entering dormancy as autumn approaches. At this time they should be dried off for the duration of winter. They should be watered lightly and occasionally towards the end of winter in preparation for their summer growing season.

Aloe dominella makes an interesting cultivation subject on account of its attractive and ornamental growth habit. It is also a very hardy species.

Right: ***Aloe dominella*** in rocky grassland near Rorke's Drift in KwaZulu-Natal. This multi-trunked specimen is typical of mature plants of this species. This illustration was painted in late September during very windy weather, pollinating insects attempting in vain to reach the flowers.

7. *Aloe ecklonis*

Aloe ecklonis Salm-Dyck

This is the most common and widespread grass aloe species in South Africa. It occurs in the grasslands of the Eastern Cape, KwaZulu-Natal, the Free State and Mpumalanga. The species also occurs in eastern Gauteng.

Aloe ecklonis is one of the few grass aloes that have adapted to grasslands extensively changed by habitat destruction, invasive alien vegetation and dense human populations. It is likely to remain in the grasslands as many other species edge toward extinction. The main reasons ensuring the success of this species concern its ability to produce enormous amounts of seed and to grow in virtually any soil type, from sand to clay. The aloes are also able to survive in habitats that have been extensively invaded by alien weeds as long as the weeds form part of the dead plant material that is burnt in winter grass fires.

The species concept of *Aloe ecklonis* was revised by Glen & Hardy (2000: 19) to include *A. boylei*, *A. hlangapies* and *A. kraussii*. This expanded concept of *A. ecklonis* has not been used in this book for various reasons which are discussed in the chapters concerning these species. Additional reasons are provided here. In terms of the knowledge construction process, those who revise plant groups follow their own rules and meanings, and operate independently from people who might have other species concepts. However, for new concepts to be accepted into general usage, other related processes have to take place from the perspective of the end user. New concepts have to be carefully articulated, not only with reference to the internal logic of the proposed model of classification, but also with reference to detailed fieldwork. These matters are fully discussed in the chapter on grass aloe taxonomy and the sociology of botanical knowledge.

Aloe ecklonis is often found in short, featureless grassland. Extensive groups of these plants are very distinctive and visible from afar, attracting a large number of nectar- and pollen-seeking insects. The result is a good seed set.

The species has been studied in the grasslands of Gauteng, KwaZulu-Natal, Mpumalanga and the Eastern Cape. The study areas all contain grassland representative of the current South African environment where these plants occur.

In eastern Gauteng and near the border with western Mpumalanga

Aloe ecklonis occurs southeast of Johannesburg in the Devon and Leandra areas. This region is extensively developed, with open-cast coal mines, maize fields and more recently expanding informal settlements. Areas not used for maize production are used for cattle farming. The grassland is heavily utilised for grazing and is degraded to varying degrees with annual weeds. Some of the remaining habitat, particularly around Devon, has been transformed by kikuyu grass (*Pennisetum clandestinum*), an introduced species from East Africa. *A. ecklonis* is rather rare in grassland used for stock farming, but prolific in road and railway-line reserves which have, until recently, excluded grazing domestic stock. The species has become particularly common along the railway-line reserve near Leandra and in the road reserve of the N17 highway near Devon.

Plants growing in road and railway-line reserves are

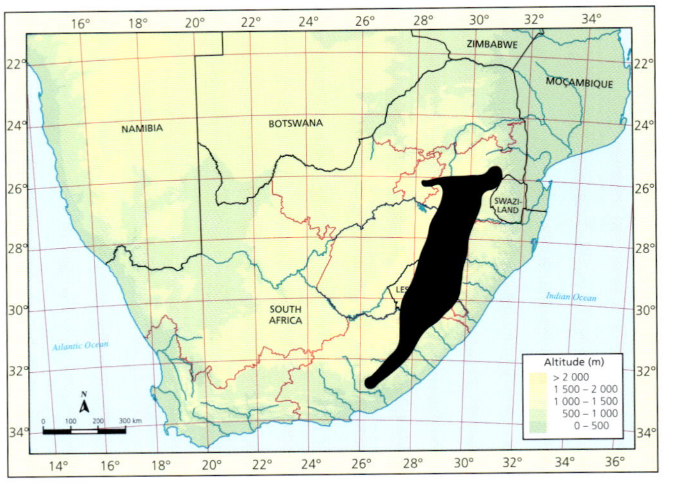

Distribution of *Aloe ecklonis*

burnt in winter veld fires. They are also mowed with grass-cutting machinery on the road verges in winter. Mowing and burning has the same effect on the plants—they flower well from one season to the next, ensuring high rates of pollination and seed dispersal. The railway-line reserves near Leandra have recently been subject to grazing by cattle and goats, leading to a sharp decline in the number of seedling plants.

The grassland north of Leandra provides a good indication of the distribution of *Aloe ecklonis* in areas that are regularly grazed. The plants are found mostly among rocks or within the shelter of large, thick tufts of grass. They are a mere fraction of the numbers of this species found in road and railway-line reserves.

Near Sheepmoor and Wakkerstroom in Mpumalanga

The Sheepmoor area lies to the west of Piet Retief and is at the transition between grassland used for stock farming and grassland that has been converted to timber plantations.

Aloe ecklonis is common in road reserves but rather scattered and less plentiful in grassland regularly used for grazing sheep and cattle. In recent decades it has become particularly abundant in an area of short grassland adjacent to the main road between Ermelo and Piet Retief where grazing animals have been excluded. This situation is directly attributable to the vast amounts of seeds set by the aloes and the fact that seedlings are not trampled by livestock. *A. ecklonis* is much less plentiful slightly further eastwards towards Piet Retief, in the timber-producing area. The aloes occur in small numbers as road reserve plants and in strips of heavily grazed grassland along streams.

In the Wakkerstroom area, the aloes are widely scattered, also growing in grassland that is regularly grazed by cattle and sheep. They often occur on rocky outcrops or among stones where they are protected from the hooves of grazing animals. In the hilly escarpment area between Wakkerstroom and Dirkiesdorp much of the habitat has been transformed by invasive black wattles (*Acacia mearnsii*).

Right: **Aloe ecklonis** growing under the boundary fence between a railway-line reserve and surrounding farmland on the Highveld near Leandra in western Mpumalanga. Railway-line reserves are important conservation areas for this species as they have, until recently, excluded grazing livestock.

In the northeastern Free State

The northeastern Free State consists of gently undulating grassland used either for maize production or stock farming. *Aloe ecklonis* is a common road reserve plant in this area. It also grows in numbers on thick clayey soil temporarily inundated in summer, sometimes for a few weeks.

Near the small town of Cornelia the plants grow together with *Haemanthus montanus*, a species that thrives in waterlogged, clayey soils. *Aloe ecklonis* is very common in the Cornelia area. It also grows on hill slopes at the edges of maize fields and may become abundant where this habitat is infrequently grazed. *A. ecklonis* habitat in the low-lying areas near Cornelia usually becomes sodden and waterlogged while the aloes are in growth and flower. Livestock prefers to graze in less waterlogged conditions on the higher ground. This serves to protect the plants from the grazing pressures characteristic of all the grassland areas where this species occurs.

In KwaZulu-Natal

The aloes are widespread and common throughout the grasslands of the KwaZulu-Natal interior. In common with habitats in other provinces, they are thinly scattered in areas where the veld is regularly grazed. A recent artificial habitat colonised by this species consists of grassy embankments at roadsides and road cuttings. *Aloe ecklonis* has established itself on the cuttings at the bottom of van Reenen's Pass where the conditions are ideally suited to its requirements. A number of young aloes are also able to take root in the soil on the cuttings, with their short, sparse grass cover.

A dwarf clump-forming *Aloe ecklonis* is found in the grassland around the foothills of the southern KwaZulu-Natal Drakensberg near Himeville. These plants conform to *A. ecklonis* in most respects apart from their clump-forming habit and dwarf flowers. *A. ecklonis* is a relatively uniform plant across its wide distribution range and anomalies such as these are likely to merit further investigation.

In the northeastern Cape

The species is well represented in much of the high-altitude grassland of the northeastern Cape. The area between Ugie and Maclear has been extensively afforested with pines. The aloes here are restricted to road reserves and rocky outcrops and are much less numerous than they are in regions where there are no timber plantations. This is related to the limited number of localities in the habitat available to these plants. Poor pollination rates appear to be indicated from the limited number of data collected in this region.

Growth cycle and the role of fire

Aloe ecklonis sprouts leaves in October and early November, at the beginning of the rainy season. The plants start to form buds in late November and early December. December is usually the peak of the flowering season, but it extends to the end of January in regions with cold climates such as the foothills of the Drakensberg in KwaZulu-Natal and the northeastern Cape. The aloes form seed capsules in January and the seeding period is usually late January and the first half of February. Seeds are dispersed during windy weather and transported to various habitats by runoff after rain.

The seeds require dry weather for effective dispersal. Large amounts go to waste if there is persistent misty or rainy weather at the time the seed capsules split. These seeds start to germinate while still in the capsules and wither when sunny weather resumes. The remaining seed stocks that have not germinated are attacked by fungi. Seed wasps may be prolific in some years, destroying a large proportion of the seed set.

Aloe ecklonis produces huge amounts of seeds and in this respect is similar to many other aloe species that do not fall within the grass aloe group. This permits the plants to respond rapidly to favourable environmental conditions, for example allowing them to recolonise grassland when grazing pressures ease.

Fires clear the grassland from moribund vegetation, particularly dense mats of dead grass. Dead leaves of *Aloe ecklonis* form large patches of flammable material in winter which is readily ignited during the course of a veld fire. Fires rid the aloe stems and leaf crowns of various pests. The flowering performance of *A. ecklonis* tapers off when the grassland has not been burnt for several years in succession.

The aloes flower equally well in grassland that has been burnt or mowed flat the previous winter. In road reserves where *Aloe ecklonis* occurs, the grassland is often mowed in winter, particularly on part of the Highveld near Devon and Leandra.

Status of wild populations

This most common and also very widely distributed grass aloe in South Africa is usually at its most abundant in road and railway-line reserves since the grasslands where this species occurs naturally, are well grazed by livestock. The view could also be taken that many large populations of this plant are artificial where they grow in areas that exclude grazing animals.

The abundance of *Aloe ecklonis* is also directly related to the increase in the number of honeybees (*Apis mellifera*). These pollinators occur commonly in much of the grassland only as a result of increased nectar sources made available through agriculture and suburban developments. Introduced insects such as spotted maize beetles (*Astylus atromaculatus*) are regularly found feeding in large numbers on the pollen of *A. ecklonis*. These insects are also likely to play some role in pollination of *A. ecklonis* flowers.

Cultivation, propagation and horticultural potential

This is an attractive, showy plant with variable flower colours—yellow, orange or even brick-red. This aloe is also very hardy and can survive in areas with severe frost and snow in winter. It grows in any soil type, from porous sands to heavy clay.

The species can be produced in large numbers from seed and as it grows in most soil types, could become a useful landscape subject. The aloes should be grown in large, deep containers, preferably in well-drained soil. Deep soil pockets in rockeries or sloping garden embankments are ideal. The aloes should be regularly watered in summer and kept dry in winter.

Seeds should be sown in any well-drained mixture in spring and early summer. They should be thinly scattered across the surface of the germination mixture and just covered with soil. Seeds are best sown in deep propagation trays or large pots. These containers should be kept moist while the seeds are germinating and for the remainder of the summer while the young plants are establishing themselves. The seedlings should be kept dry for the duration of the winter except for an occasional watering in the late winter and early spring, before the beginning of the growing season. Seedlings are usually well established with a large root system by the end of the second growing season. At this stage they should be transplanted.

8. *Aloe fouriei*

Aloe fouriei D.S.Hardy & Glen

This is one of three grass aloe species that are similar to one another with respect to morphology and habitat preference. The others are *Aloe verecunda* and *A. vossii*. A taxonomic re-evaluation of these plants is required. The need for such a reappraisal has also been recognised by Andrew Hankey, senior horticulturist at the Walter Sisulu National Botanical Garden, and Pieter Winter, taxonomist at the South African National Biodiversity Institute. Between them, these two botanists have substantial fieldwork, herbarium and horticultural experience with the three grass aloe species, and the author has spent many years collecting data from plants in the different populations of these species.

The consensus of opinion is that, for the three species to be retained, *Aloe verecunda* has to be circumscribed with clearly defined boundaries that separate it from *A. fouriei* and *A. vossii*. This will require detailed sampling of plants from many populations, including the type locality from which *A. verecunda* was collected at Ashmoledales in the Wolkberg.

The problems with the circumscription of the three species in the knowledge construction process is reflected in the herbarium record. Plants collected from the dolomites of the Drakensberg near the Abel Erasmus Pass in Limpopo Province on 11 February 1969 were recorded in the National Herbarium as a form of *Aloe verecunda*. This diagnosis was later changed to *A. fouriei* once this species had been described. A specimen of *A. verecunda* in the National Herbarium collected by aloe expert the late David Hardy on 16 November 1976 at The Downs, also in the Drakensberg, was said to have affinities with *A. verecunda* and *A. vossii*. Another *A. verecunda* herbarium specimen from The Downs, also in the National Herbarium, appears to have the characteristic flowers of the entity now known as *A. fouriei*. This specimen was collected on 12 March 1986.

Further research may indicate that *Aloe fouriei* and *A. vossii* are ecotypes of *A. verecunda*. The Downs and the surrounding areas would certainly be a good place to begin studying the plants, with a view to collecting data that can be used towards reassessing the taxonomic position of these plants. The three names are retained in this work with the full recognition that the correct taxonomic placement of these entities is far from settled.

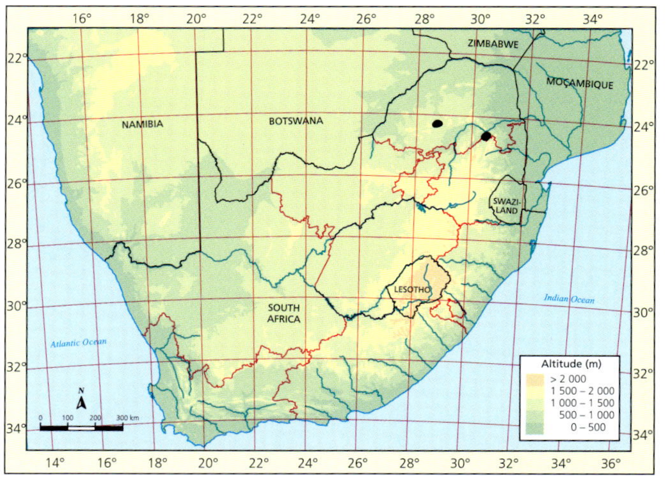

Distribution of *Aloe fouriei*

Plants that conform to the type description of *Aloe fouriei* are found over a wide geographical area in Limpopo Province. Typical plants occur on the dolomites of the Abel Erasmus Pass where much of the data for this chapter were collected. Further populations have been found in the Dwars River Pass to the southwest. Two additional populations of what appears to be this species have been found in the Highlands Mountains southwest of Polokwane and on the summit of the Strydpoort Mountains to the east of Chuniespoort. Chuniespoort is not far, as the crow flies, from the Highlands Mountains. Darrel Plowes has also commented on an *Aloe* species close to *A. verecunda* which he photographed on the summit of a mountain east of Chuniespoort. A photograph of the plant appears in the journal *Aloe* (22,3: 69, 1985) and seems to represent what is now known as *A. fouriei*.

Aloe fouriei grows on dolomite around the Abel Erasmus Pass, on the Highlands Mountains and on the peaks of the Strydpoort Mountains east of Chuniespoort. The plants are also found on norite, in the Dwars River Pass region. This indicates that the plants are not endemic to dolomite substrates and this will have to be taken into account when this species, *A. verecunda* and *A. vossii* are reassessed.

The plants have been studied on the dolomites of the Abel Erasmus Pass, in the Dwars River Pass area, on the Highlands Mountains and on the peaks of the Strydpoort Mountains to the east of Chuniespoort. The most detailed observations have been made in the vicinity of the Abel Erasmus Pass over a period of about a decade.

Near the Abel Erasmus Pass

Aloe fouriei is well represented on some of the dolomite mountains in the vicinity of the Abel Erasmus Pass south of the Drakensberg Escarpment. The plants are found in several situations—in clumps at the edges of large sheets of exposed dolomite usually amongst rocks, on cliff faces, and amongst tumbled rocks on the summits of large hills.

The aloes grow mainly on east- or south-facing slopes. They receive morning sunlight on the eastern slopes but are shaded for much of the day where they occur in south-facing positions, particularly cliff faces that face due south. Plants growing on the summit of the hills are exposed to day-long sunlight and as a result are found in positions that retain the greatest amount of moisture after rain. The habitat is arid for a grass aloe and the plants occur together with various species of *Euphorbia* and *Crassula* and bulbs such as *Ledebouria zebrine* adapted to arid conditions.

Plants growing in exposed, arid situations tend to be short and clump-forming. They often occur in large mats, with many stems. Those growing in shaded positions, particularly on ledges of south-facing cliffs, are usually robust, with small numbers of individual stems. It is surprising to find a grass aloe that is relatively numerous under rather arid conditions in countryside subject to frequent and prolonged droughts. The plants survive by growing in positions that are very well irrigated when it rains and retain moisture for several days afterwards. Dolomite-derived soil furthermore retains water and drains slowly.

The habitat has become increasingly degraded with the passage of time. Goats and sheep kept by local pastoralists graze the slopes heavily throughout the year and firewood is removed from stands of indigenous trees. Trampling by grazing livestock has seen the demise of some clumps of aloes growing in exposed positions beside sheets of dolomite. *Aloe fouriei* is now found mostly wedged amongst rocks and on cliff faces where the plants are not trampled by grazing animals.

Plants of *Aloe fouriei* near the Abel Erasmus Pass has a very long flowering period, extending from early November to early March. Some plants usually come into flower after a good shower of rain. Flowering is usually sparse and seed production consequently limited. Regeneration from seed is slow and most of the aloes occur in large, well established clumps. No seedlings have been noted in recent years in the areas subject to heavy grazing.

In the Dwars River Pass

The species has a curious distribution in the area surrounding the Dwars River Pass. The aloes are found in valleys beside seasonal streams amongst tumbled boulders and also on rocky mountainsides. Individual plants are frequently surrounded by tall grass during successive years when the veld has not been burnt in winter grass fires. The plants are rather sparse in areas with the densest grass cover since they rarely flower in these situations and seed is set rather infrequently. At present, the Dwars River Pass area represents the southern distribution range of *Aloe fouriei*. The area between the Dwars River Pass and Burgersfort has not been searched for this species since it is rather remote and inaccessible. If *A. fouriei* is found in the mountains in this region and also between Burgersfort and the Abel Erasmus Pass, it may provide one reason in favour of upholding the specific rank of *A. fouriei*.

On the Highlands Mountains and Strydpoort Mountains east of Chuniespoort

Fanie Venter, well-known and widely experienced field botanist, found a small clump-forming grass aloe growing on dolomite in the Highlands Mountains southwest of Polokwane. The plants were found while Fanie was engaged in a botanical survey for the Klipspringer Diamond Mine, which has recently been established in the southern foothills of the Highlands Mountains. A field visit took place in early April 1998 at a time when the grass aloes were starting to enter their winter dormancy period. Fanie had obtained special permission from the management of the mine for the author and Andrew Hankey to accompany him.

The aloes were found amongst porous dolomite rocks at the edge of a small ridge of low cliffs. The plants formed large mats of closely packed stems, mostly only about 8–12 cm high. The growth habit of these plants and their diminutive stature at first conveyed the impression of an undescribed species. It subsequently became apparent from plants in cultivation that these grass aloes conformed to most of the characters described for *Aloe fouriei*. The habitat is rather arid and the aloes were found only on southern and southeastern slopes, growing amongst rocks and tufts of short grass. The hottest slopes and rocky summits of hills exposed to day-long sunlight contained a rich

Right: ***Aloe fouriei*** growing in the crevice of a low dolomite cliff on the mountains near the Abel Erasmus Pass in Limpopo Province. South-facing cliffs are shaded for much of the day and retain moisture the longest after rain, offering ideal habitat for the species.

succulent flora, including a large colony of *Euphorbia restricta*, perhaps the furthest west that this species occurs, but the aloes were not found in these arid situations.

The Strydpoort Mountains immediately east of Chuniespoort are very steep and clad with dense indigenous trees. The summits of the mountain peaks contain a limited amount of exposed, rocky grassland. The same dwarf *Aloe fouriei* found in the southern foothills of the Highlands Mountains also occur in this area, but more sparsely. They share their habitat with *Ledebouria dolomiticola*, a species with a partly exposed bulb, also partial to porous dolomite. This region is difficult to survey since it is extremely rugged and remote. It can be expected, though, that *A. fouriei* is to be found right along the Strydpoort Mountains. The eastern boundary of the Strydpoort Mountains is not far from Ashmoledales, the type locality of *A. verecunda*. It would be important for future researchers to look carefully at plants of both species in this area when circumscribing the species concept of *A. verecunda*.

Growth cycle and the role of fire

Aloe fouriei grows from early November until late February or early March, the time of year when most of the rainfall can be expected. The plants flower at any time during this period after a good shower of rain. The flowering period is much longer than that of most grass aloes, but it is rare to find many plants in flower simultaneously at one locality. Occasionally prolonged wet weather followed by sunny spells promotes mass flowering. These conditions are rare, usually occurring about once a decade in the more arid habitats such as that around the Abel Erasmus Pass. There are rarely enough plants flowering simultaneously to attract large numbers of pollinators. It is for this reason that seedlings are rarely noticed in habitats where *A. fouriei* is encountered.

The aloes enter a state of semidormancy during prolonged droughts, but they quickly resume growth again even after one light rain shower. Light precipitation is sufficient to irrigate the soil thoroughly in habitats where the water is funnelled off large sheets of exposed dolomite onto the surrounding ground.

Fire is significant in all areas where *Aloe fouriei* is found, clearing the habitat of dead grass. Heavy grazing around the Abel Erasmus Pass accounts for light burns owing to the limited amount of dry grass during the dry winter months. In former times when fires were more frequent, the large groups of *A. fouriei* were in a more healthy state than they are currently. Dead leaves and dry stem material were burnt away, cleaning the plants from parasites and fungal spores.

Status of wild populations

Plants that conform to the currently accepted circumscription of *Aloe fouriei* are fairly widespread. Those on the Strydpoort Mountains are conserved by virtue of the remoteness and ruggedness of the terrain, which is unsuited to pastoralism. Aloes in the Dwars River Pass area have been destroyed by mining activities, but this is insignificant as mining poses few problems for this species overall.

The most extensive populations of plants are currently found in the Abel Erasmus Pass area, which is densely settled. Heavy grazing pressures are likely to lead to increased desiccation of the already arid mountain slopes where *A. fouriei* grows. In the long term this is probably more significant than trampling by livestock since only plants growing out in the open beside sheets of exposed rock are exposed to trampling.

Cultivation, propagation and horticultural potential

Aloe fouriei grows well in cultivation, either on sunny rockeries or in containers. Earthenware pots with a rough finish are the most suitable for cultivation since they drain more rapidly than plastic containers. Plants should be grown in a well-drained medium consisting of one third gravel, one third coarse sand and one third ordinary sandy soil to which some well-rotted leaf mould and very well-rotted kraal manure have been added. The aloes should be exposed to half a day's direct sunlight and grown in positions where there is plenty of free air circulation.

This species roots very readily from cuttings. Stems should be broken off the parent plants in late winter or early spring. They should be broken off rather than cut off with a blade, since the healed break allows a greater surface area for development of roots. Broken stems should be left to harden off in a cool, dry place before planting. The stems should be covered to a depth of about 2–3 cm and kept just moist until they are well rooted. Plants should be rooted in gravel and then transferred to the pots or sites in the garden where they will be permanently cultivated.

Aloe fouriei is also readily grown from seed but so little is set in cultivation, even with hand pollination, that propagation from cuttings is usually the preferred method. Seeds should be lightly and evenly scattered across the sowing medium and just covered. Large, deep propagation trays and deep pots are the best containers to use when propagating the plants from seed. The seed trays should be kept moist during germination and for a few months afterwards until the seedlings establish their first set of leaves. The young plants are fairly hardy at this stage and can be watered less frequently.

This species makes a very attractive container subject, particularly when planted in terracotta containers with a rough finish. The aloe stems and leaves can be effectively displayed amongst dark, lichen-blotched, upright rocks such as fragments of dolomite. The plants flower rather irregularly and may skip a season or two without flowering when they are grown in containers. The paucity of flowers is a natural habit which does not appear to be altered much under cultivated conditions when plants are grown relatively hard. Care has to be taken to ensure that the aloes receive enough direct sunlight to preserve their compact natural growth form.

Aloe fouriei is regularly attacked by root mealybug and woolly aphids. These pests are particularly prevalent during hot, dry weather.

9. *Aloe hlangapies*

Aloe hlangapies Groenew.

Aloe hlangapies first became known to the botanical world from the Piet Retief District of Mpumalanga where the typical plants still occur. They are generally characterised by distichous leaves and densely capitate racemes of apricot-coloured flowers. The accompanying illustrations were drawn and painted in the field southwest and east of Piet Retief more than 50 years after this aloe had been described. The plants have been all but eliminated from much of their former habitat by extensive afforestation with alien eucalypts and pines.

The similarity of *Aloe hlangapies* to *A. ecklonis* has been the subject of comment and discussion from 1938 shortly after the species was formally described, until 1996 (Pole Evans 1938: Plate 710; Reynolds 1969: 137–139; Van Wyk & Smith 1996: 264–265). *A. hlangapies* was reduced to synonymy by Glen & Hardy (2000: 19–20) in their revision of the genus *Aloe*.

However, the species concept of *Aloe hlangapies* as proposed by Reynolds is upheld in this work for several reasons. At the theoretical level of knowledge construction, where various species are reduced to synonymy under *A. ecklonis*, no reasons for this decision are advanced by the authors. One of the critical dimensions for knowledge construction in *Aloe* taxonomy is that reasons are advanced for these types of decisions. In the absence of such reasons, species concepts in *Aloe* cannot be adequately articulated. There is consequently a reversion from the perspective of the sociology of botanical knowledge, to the previous paradigm until there are further theoretical developments.

Several authors have referred to *Aloe hlangapies* crossing and intergrading with *A. ecklonis* (Pole Evans 1938: Plate 710; Reynolds 1969: 138; Van Wyk & Smith 1996: 264). These intergrades and crosses have never been accurately diagnosed, however. It is not clear whether they are natural hybrids between two species or manifestations of variation in one species across different geographical areas. The fieldwork for this book took place at a time when the Piet Retief District was already extensively afforested with timber trees. *A. hlangapies* occurs in scattered groups in small areas of habitat between timber plantations and in road reserves. The few populations of plants where the two species appear to intergrade with one another are

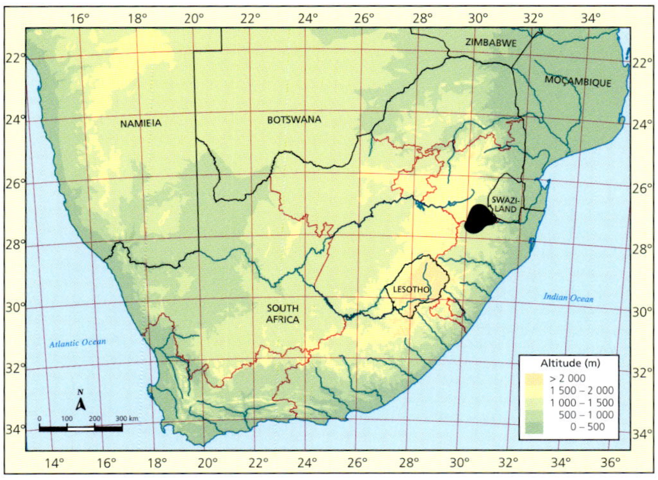

Distribution of *Aloe hlangapies*

apparently hybrids. These hybrid populations have been brought about by changes to the pollination biology of the two aloes. This process is discussed in more detail below, where details are given of plants growing in specific colonies near Piet Retief.

The most significant reason in favour of upholding *Aloe hlangapies* as distinct from *A. ecklonis*, are the different flowering seasons. *A. hlangapies* flowers from mid-October to mid-November, with the peak flowering season usually in the last week of October and first week of November. *A. ecklonis* flowers mostly in late November and December in the districts where *A. hlangapies* also occurs. The chances for hybridisation and intergrading as it has been called, are only possible where the latest flowering plants of *A. hlangapies* occur close to the earliest flowering ones of *A. ecklonis*. Hybrid populations contain plants that flower at any time between mid-October and early December.

The modern environment near Piet Retief is heavily afforested with timber plantations and the *Aloe hlangapies* populations are very fragmented. As a result, detailed sampling of *A. hlangapies* and *A. ecklonis* is unlikely to yield reliable information on variation in the populations of these two species.

One of the main factors said to separate *Aloe hlangapies* from *A. ecklonis* is the distichous leaves of the former. This character is not constant, however, and some plants of *A. hlangapies* have rosulate leaves. The leaves and the densely capitate racemes of apricot-coloured flowers are the main characters distinguishing *A. hlangapies* from *A. ecklonis*. To some this may hardly justify separation at species level, but if the earlier flowering time of *A. hlangapies* is taken into account, this additional factor strengthens the case for keeping the two entities separate.

Aloe hlangapies has been studied at a number of localities, the Dirkiesdorp, Luneberg, Paulpietersburg and Piet Retief Districts. The greater proportion of this region is used for timber production except for the mountainous area around Luneberg which still contains tracts of mountainous grassland, albeit heavily invaded with black wattles (*Acacia mearnsii*).

Road reserve colonies near Piet Retief

Most plants of *Aloe hlangapies* are found in small road reserve colonies at the edges of eucalypt and pine plantations, usually in rather small groups of five to ten individual plants. The grassland in the road reserves is either burnt as firebreaks every year or mowed as part of road maintenance activities. Most plants in these groups consist of old multibranched specimens that are the remnants of extensive colonies of the species that used to occur in the grassland before it was afforested.

The aloes flower well each year owing to the clearing of the habitat, mostly through fire. The seed set is usually very poor owing to the lack of pollinators in road reserves surrounded by timber plantations. Few seedlings are generally observed and where they do occur they are usually found amongst rocks where they escape the worst effects of burning.

Some of the plants in road reserves occur where there are steep cuttings. This man-made habitat sometimes has large numbers of aloes where the ground is porous and where short tufts of grass have also become established. Plants with characters of both *Aloe hlangapies* and *A. ecklonis* are sometimes found on road cutting embankments.

The flowers and ripening fruit of the natural hybrid between **Aloe hlangapies** and **A. ecklonis** send out signals to the trained field observer of both species. These flowers and fruit are from the same locality near Piet Retief in Mpumalanga where the hybrid is depicted in bud (page 60).

They appear to be hybrids produced when *A. hlangapies* and *A. ecklonis* overlap in their flowering seasons and where they grow in close proximity to one another. Where these plants grow close together on embankments, but are otherwise hemmed in by plantations, they sometimes attract sunbird pollinators (species of *Nectarinia*). Sunbirds are very effective pollinators of many aloe species and they increasingly visit mixed *A. hlangapies* populations if they expand in size along the artificial road cutting habitat. The sunbirds repeatedly visit the same flowering plants during pollination episodes, where large numbers of aloes are in flower in a confined area. This results in some populations of *A. hlangapies* that have attributes of *A. ecklonis*, particularly with respect to leaf arrangement and flower colour.

Populations in hilly grassland near Piet Retief and Amsterdam

Some areas of undulating grassland close to Piet Retief have escaped the development of timber plantations. Other areas northwest of Piet Retief, not far from the small town of Amsterdam, consist of rocky, hilly grassland harbouring sheets of exposed rock. These are unsuitable for the establishment of eucalypt and pine plantations and are the remains of what was once a much more extensive habitat. These areas still support fragmented colonies of *Aloe hlangapies*.

Very robust examples of *Aloe hlangapies* occur east of Piet Retief. The plants at this locality are typical examples of the concept of *A. hlangapies* used by Reynolds. They have large, densely capitate racemes of apricot-coloured flowers and distinctively broad, distichous leaves. The aloes grow in deep soil close to seasonally moist, sloping seepage areas that drain into a stream in the valley below. The habitat is burnt nearly every winter and flowering is consequently prolific from one year to the next. Despite the frequency of fires, a proportion of young plants of various ages are found. This is related to the dense grass cover in this habitat and the fact that large tufts of grass are not burnt right to the ground in veld fires. By the time the young plants are a few centimetres tall they are well able to withstand even the hottest burns during the course of a grass fire.

This colony of plants lies within an area that experiences hailstorms in the early summer. Every few years, flowers and ripening seed capsules are completely destroyed during strong hailstorms.

Aloe hlangapies is at the northernmost section of its distribution range near Amsterdam. Further north the plants are typically *A. ecklonis*. *A. hlangapies* is common on an extensive rocky hill southwest of Amsterdam. The plants are rather dwarf in stature, less than a third of the size of those found close to Piet Retief. They grow at the edge of sheets of sloping rock and amongst tumbled boulders, but are absent in a narrow strip of riparian grassland with deep soil in the valley that is subject to frequent flooding in summer.

Right: The typical form of **Aloe hlangapies** with distichous leaves and rich apricot-coloured flowers borne in a capitate raceme. This plant was found in a strip of riparian grassland on the outskirts of Piet Retief in Mpumalanga.

J Condy

The aloes on the rocky hillside tend to sucker repeatedly and sometimes form large, multistemmed clumps. Flowering performance is best after a winter grass fire, but fires are actively discouraged since the habitat lies within a fire prevention zone.

Near Dirkiesdorp, Luneberg and Paulpietersburg

Dirkiesdorp, Luneberg and Paulpietersburg mark roughly the western, southwestern and southern boundaries of the *Aloe hlangapies* distribution range. A few outlying populations are occasionally encountered. Outside this area the typical *A. ecklonis* plants are found, particularly around Wakkerstroom and Volksrust.

The grassland in these districts is transformed and degraded to varying degrees. Much of it is under newly planted timber. In the last two decades a number of open-cast coal mines have been established in the Dirkiesdorp area and a more limited amount of grassland in this region and near Luneberg has been put under the plough. *Aloe hlangapies* occurs in small isolated populations in these districts, but on the whole *A. ecklonis* is more numerous.

The Dirkiesdorp area, until recently, would have provided an ideal region from which to sample various *Aloe ecklonis* and *A. hlangapies* populations. Data collected in the grasslands around Dirkiesdorp would certainly have served the useful purpose of contributing towards the debate on the attributes that separate the two species. The *A. hlangapies* habitat around Dirkiesdorp has now become too fragmented by afforestation, mining and serious invasion of the grassland by alien trees for meaningful sampling.

Growth cycle and the role of fire

Aloe hlangapies develops its leaves mostly from mid-September to mid-October. The earliest flowering plants are usually found in veld that has been burnt the previous winter. The emerging flower buds are often evident by mid- to late September in burnt grassland, but only appear about three weeks later in areas that escaped winter fires.

The plants are in full flower from late October to mid-November. Seeds are liberated into the environment during the second half of December and up to about mid-January in the case of the plants that flowered late. Seeds are dispersed mostly by wind and water runoff after rain. Much of the seed germinates in late January and February if there is enough consistent rain to trigger germination.

Fire is essential for mass flowering of this species and the flowering performance of these aloes tapers off if there have been no grass fires for several years in succession. The charred thickened aloe stems are effective repellents of fungal attack. This species accumulates large amounts of dead leaf material harbouring fungal spores and in this respect is similar to all other large grass aloe species.

Aloe hlangapies starts to enter dormancy between early March and late April. This depends very much on the end of the rainy season. In years when the rains are late to finish, the leaves remain greenest for longest. The plants rapidly enter full dormancy after early frosts, which can be expected during the first three weeks of May.

Left: The natural hybrid between ***Aloe hlangapies*** and ***A. ecklonis*** in bud near Piet Retief in Mpumalanga. It occurs in places where the two species grow together and there is some overlap in their flowering time.

Status of wild populations

Aloe hlangapies exists in very fragmented habitats on the whole, some of them removed from natural fire cycles and pollination processes. Many of the road reserve populations do not have a secure long-term future. These areas are used for grazing livestock and are ploughed up in some areas. Habitat degradation and transformation by introduced alien vegetation is a severe problem in some areas, particularly southwest of Dirkiesdorp.

The status of *Aloe hlangapies* remains uncertain. If it is ever adequately reduced to synonymy under *A. ecklonis*, it is little more than a variant of this species, the most common and most widespread grass aloe in South Africa. However, this problem can be adequately resolved only by extensive sampling of *A. hlangapies* and *A. ecklonis* in the same geographical area. It is doubtful whether this is possible owing to severe habitat loss and fragmentation, which would inevitably bias the results of a detailed survey.

In the early days of botanical exploration in South Africa the lack of road networks made population studies of aloes a difficult task. This is no longer a problem, but populations of many plant species fragmented by habitat destruction will have to be given serious consideration by plant taxonomists in the future.

Cultivation, propagation and horticultural potential

Aloe hlangapies is a very showy species easily propagated from seeds, suckers and stem cuttings but difficult to grow well in cultivation. Plants are prone to attack by root mealybug when they are grown in containers. They also flower somewhat erratically.

The species should be grown in large, deep containers in a mixture of equal parts coarse gravelly sand, sieved sandy soil and very well-rotted leaf compost. An alternative to leaf compost is kraal manure that has been allowed to decompose in the sun. The aloes grow best where they are exposed to about half a day's direct sunlight. They also favour positions of very lightly dappled shade. The plants should be watered thoroughly from late spring to early autumn and kept largely dry during the winter months. They should be lightly and occasionally watered during late winter and early spring.

This species is ideally suited to growing in patches of natural or artificially created grassland that is burnt every second or third year, towards the end of winter.

The plants set some seed with hand pollination. Seeds should be lightly and evenly scattered across the surface of the soil and just covered. They should be well watered but not kept too damp during germination and until the beginning of dormancy. Water should gradually be withheld as autumn progresses to winter. Large pots are best for propagation from seed, since it allows the plants to develop their extensive root system. This practice also avoids the necessity of transplanting the seedlings into bigger containers as they mature.

Cuttings and suckers should be rooted in fine gravel in spring or early autumn when roots will develop rapidly, usually within four to six weeks. Rooted cuttings can be kept in containers for several months and have to be planted only the following spring if rooted in the early autumn.

10. *Aloe inconspicua*

Aloe inconspicua Plowes

Aloe inconspicua is one of three bulbous grass aloe species found in South Africa. The plants are known only from a small area of hilly thornveld east of Estcourt in KwaZulu-Natal. They are found mostly in short grassland, generally growing on gently sloping ground beside large hills. Most records are from farms in the vicinity of the Bushman's River.

This remarkable grass aloe was discovered by David Green, a farmer from Estcourt, in December 1972. The plants had seed capsules at that stage. These capsules and the thin, linear, succulent leaves with small teeth placed the plants provisionally in the genus *Aloe* but it was not until about a year later, when flowers were observed, that this could be confirmed.

Green could not place the aloe after consulting Barbara Jeppe's book *South African aloes* (1969). He started correspondence about the aloe with various well-known South African botanists and sent them material for identification. A perusal of this correspondence between 1975 and 1986 is very interesting. The botanists involved were Olive Hilliard at the University of Natal and the late Amelia Obermeyer Mauve (Mrs Mauve) of the then Botanical Research Institute in Pretoria. The late Lynette Davidson of the Botany Department of the University of the Witwatersrand and the late David Hardy of the Botanical Research Institute also expressed their views on the correct placement of the plant during the course of the correspondence.

Olive Hilliard initially thought that the plant was probably a species of *Chortolirion*, as discussed in her letter to David Green dated 14 December 1973. She also sent some dried material to Jim Ross at Kew. In a subsequent letter sent to Green, 30 April 1975, Hilliard indicated that the plant could not be matched with existing material at Kew. A herbarium specimen was subsequently forwarded to Mrs Mauve at the Botanical Research Institute in Pretoria who consulted David Hardy, the resident *Aloe* expert at the time. He immediately identified it as *A. myriacantha* and Mrs Mauve conveyed this species identification to Olive Hilliard in a letter dated 29 October 1976, who in turn forwarded the letter to David Green on 12 November 1976.

Lynette Davidson of the Moss Herbarium at the University of the Witwatersrand in Johannesburg independently identified the plants sent to her as *Haworthia angolensis* (*Chortolirion angolense*). This was in October 1979.

David Green discussed his find with Godfrey Symons a well-known naturalist with a thornveld farm close by. Symons informed his cousin Darrel Plowes whom Green subsequently took to the first known colony. The species was finally described as *Aloe inconspicua* by Plowes in 1986. In his description, Plowes was of the opinion that this species did not readily convey details leading to its placement in *Aloe*. This was so as a result of the sessile attachment of the flowers and the bulb, which is rarely found in species of *Aloe*. However, he drew attention to the characters that led David Green to believe that the species belonged to the genus *Aloe*, namely the marginal teeth on the leaf surfaces as well as the capsules and seeds. In a concluding letter to the correspondence between Olive Hilliard and David Green, she congratulated him on his persistence to get the plant correctly named.

An analysis of the correspondence is instructive as far as the production of botanical knowledge is concerned. Some of the factors leading to different perceptions of where *Aloe inconspicua* should be placed, are discussed below. There is little difference between the bulbs and leaves of *A. inconspicua* and those of some broad-leaved examples of *Chortolirion angolense*. In the absence of flowers the two can be confused, and even the capsules of the two entities have some similarities—until the seeds are examined. *A. myriacantha* does not usually produce a pronounced bulb-like swelling similar to that of *A. inconspicua*. Some examples of *A. myriacantha* do, however, have bulb-like swellings and these plants are closely resemble plants of *A. inconspicua* with slender bulbs.

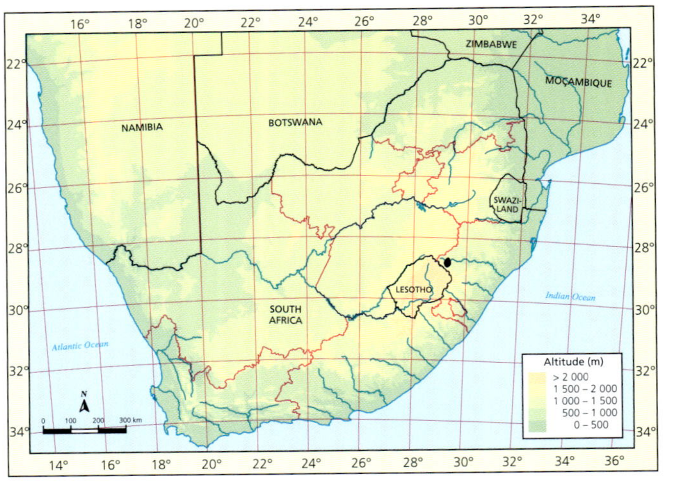

Distribution of *Aloe inconspicua*

All the botanists involved were working with their perceptions of characters that defined the genus *Aloe* and genus *Chortolirion*. The placement of *A. inconspicua* required the examination of several characters as a whole. This is precisely what Plowes (1986: 32–33) has done in his short but concise description of the species.

Aloe inconspicua differs in terms of its autecology from other grass aloe species. If this dimension is studied closely, the species is an anomaly correctly placed in the genus *Aloe*, but with obscure affinities. The text that follows clarifies these aspects in more detail.

Aloe inconspicua has been studied at two localities, both east of Estcourt in hilly grassland amongst acacias. In the one area just east of Estcourt much of the veld is now part of an informal settlement. The other area is on a farm where the grassland is grazed by cattle. The countryside is well cared for by the farmer who has a particular interest in the preservation of rare indigenous flora.

The future of *Aloe inconspicua* in its currently known distribution range east of Estcourt is uncertain. The farmers have been campaigning to have their land incorporated into one large game reserve. This initiative towards an alternative to cattle farming is the result of consistent stock theft. In the medium to long term these circumstances are likely to lead to the demise of cattle ranching in the area. The matter is a long way from settled and it is possible that

Right: Two robust specimens of **Aloe inconspicua** in flower and in seed. The rich green seeds, which ripen from late December to January, are often more conspicuous than the flowers. The tip of the inflorescence often withers if the weather is very hot and dry during November when the plants flower. These plants characteristically produce a few basal seed capsules.

J Condy

Aloe inconspicua in a road reserve near Estcourt in KwaZulu-Natal. Plants occurring in situations like this with stony, often heavily compacted soil are depauperate in stature. Those growing in deeper soil in less arid grassy places are often much larger, producing more flowers and therefore greater volumes of seed.

other forms of land use, such as small-scale stock farming, may be contemplated. Some indications of how the plants would adjust to this form of land use have already been provided from data collected in the vicinity of the informal settlement.

Around the outskirts of the informal settlement

Aloe inconspicua used to be well represented in hilly grassland thornveld immediately east of Estcourt. The plants occupied all the available habitats where the species is usually found. This includes pebbly stony ground, open patches of shaly soil and patches of short grass at the base of steep hills. The veld was well preserved, in common with much grassland on the outskirts of many towns in eastern South Africa. Dumping of garden refuse was a minor problem as this took place where the smallest numbers of *A. inconspicua* were to be found. Numbers of young plants were observed, indicating a healthy recruitment of seedlings to the different groups of plants throughout the habitat.

A large informal settlement has recently been built, eliminating a good proportion of former groups of *Aloe inconspicua* in the area. The aloes still occur on the eastern outskirts of the settlement. This area is heavily grazed by cattle, goats and sheep and firewood is regularly collected from the trees. The ground has become progressively compacted as a result of over-grazing and few young plants are currently observed. Most aloes are now found in stony ground or on open patches of shaly soil exposed to strong direct sunlight for most of the day. These situations are the least attractive to grazing animals and the aloes growing amongst rocks are least susceptible to trampling.

The grassland surrounding the informal settlement is regularly burnt in winter and areas that escape burning are kept short by grazing livestock. These conditions are conducive to good flowering and seeding, but the possibilities for regeneration are limited to rocky areas where young plants escape trampling.

It is possible that felling trees for firewood may increase the habitat available to *Aloe inconspicua*. Trees are often found in stony areas with some pockets of deep soil, a habitat type widely used by *A. inconspicua* at this locality.

Colonies on a cattle ranch east of Estcourt

Aloe inconspicua has thrived on a farm east of Estcourt for many years. The veld is well managed and regularly but lightly grazed. The species is most common in open grassy patches, usually where the grass cover is at its thinnest. In this situation the aloes occur in groups of mature bulbs and seedlings of various ages. The veld is rarely burnt since it is kept short by grazing cattle, with little need for burning out moribund grass and herbs.

There is a large amount of suitable habitat in the general area but only small sections of it have been used by the aloes. The plants were also not found on old lands used for crop farming that had been allowed to revert to the natural vegetation. The various groups of aloes flower reliably each year in November, followed by a good seed set in late December and January.

Aloe inconspicua is more numerous at this locality than it is immediately east of Estcourt. Before establishment of the informal settlement the plants grew in grassland that became very dense and obtained the chance to seed well, with conditions suitable for good germination, only in summers after a grass fire the previous winter. It is evident that *A. inconspicua* is commensal with regular but light grazing, circumstances evident on the cattle ranch.

Growth cycle and the role of fire

Aloe inconspicua starts its growth cycle in October with the beginning of warm late spring weather. The plants have produced their set of annual leaves by November. The flowering season is invariably the first three weeks of November, with occasional plants flowering in the last week of November or the first week of December. The flowering season may or may not coincide with the first rains of summer. This species is not dependent on rainfall to initiate growth and flowering. The species often flowers under hot, dry, rainless conditions.

The conspicuous dark green seed capsules start to form in late November and are usually fully developed by mid-December. The plants are at their most visible at this time of year. Seeds are usually shed in January, but some early flowering plants are in seed just before the southern summer equinox, 22 December. Few seeds are produced in comparison with most other grass aloes. A total of 15 randomly selected plants with seed capsules in the population immediately east of Estcourt were examined in early December 2004. The fewest number of seed capsules per plant was three and the greatest number 13, with an average of seven capsules per plant. One plant in the process of shedding seeds had four capsules. The greatest number of seeds in a capsule was 17 and the fewest nine, with an average of 12 seeds per capsule. The total number of seeds totalled 49, most parasitised by seed wasps.

The number of seeds liberated into the environment is always low and seriously impaired in years when they are heavily parasitised by the seed wasps. There is some compensation, however. Fertile seeds are very hard and triangular in shape. They are not as readily blown about in the wind as the flatter, light seeds of most grass aloes. Seeds are dispersed mostly by water runoff after rain, which ensures that many of them land on soft soil in areas suitable for germination. Only some of the seed germinates the first year it is liberated and the remainder the following summer.

The clayey soil where *Aloe inconspicua* grows frequently cracks during dry weather. Some of the seed is lost when it falls into fissures and lodges at too great a depth under the soil for it to germinate. The best seed dispersal conditions are provided by regular rainfall in January which ensures that the clayey soil does not crack open too much during spells of dry weather.

Aloe inconspicua currently has limited dependence on fire cycles for flowering or seeding. Many of the plants grow on open clayey soil where there is very little flammable material in the dry winter months. The advent of cattle farming has ensured that the veld is regularly grazed and kept short. A few groups of *A. inconspicua* are found at the base of stony hillsides where the grass grows long. In these situations fire is necessary to clear the habitat periodically. The flowering performance of these aloes is improved after fire the previous winter. Fire is likely to have been more significant in former times before cattle ranching. It is unlikely, however, to have been integrally related to the autecology of the plants since they grow mostly in very open situations in strong direct sunlight, often with minimal grass cover.

Status of wild populations

Aloe inconspicua is well represented in a few isolated localities east of Estcourt where cattle ranching is the main form of land use. The plants are, however, absent from a large part of the countryside that appears ideally suited to their requirements.

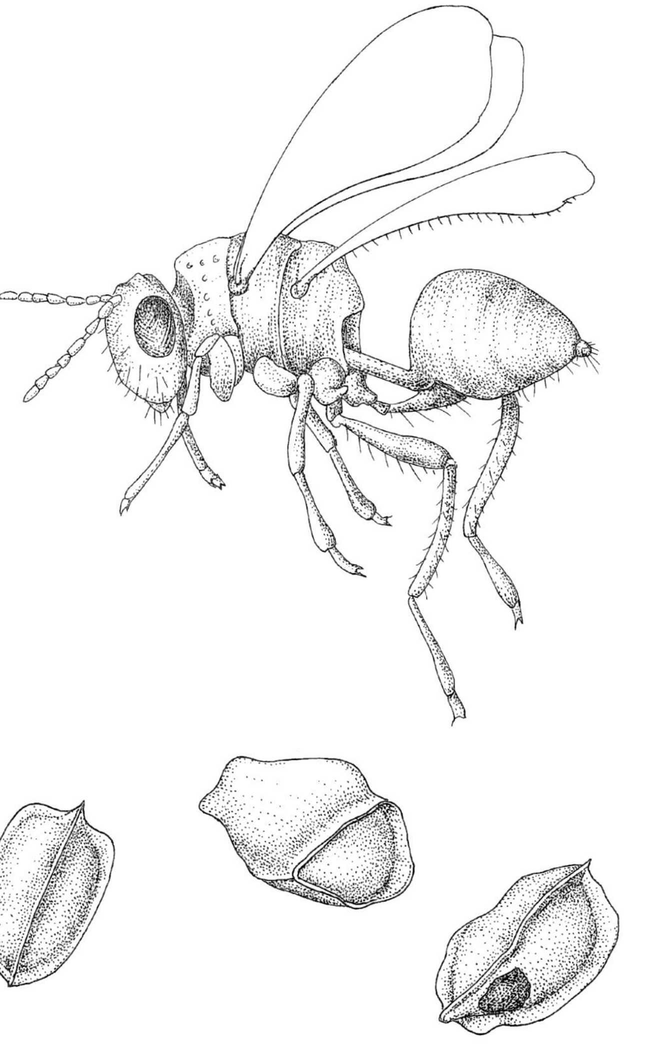

One of the seed chalcids (species unidentified) that parasitise the seeds of ***Aloe inconspicua***, and seeds of this species, all magnified 50 times. Seeds of this aloe are sometimes very heavily parasitised by seed wasps in its habitat near Estcourt in KwaZulu-Natal. The seed on the right shows the characteristic hole indicative of this parasitism. It is likely that several species of these wasps are undescribed.

Changes to land use in the area are likely to affect the status of the plants. Should the area be used for a game reserve, the future of the plants would be secured. If, on the other hand, the farms are converted to areas for indigenous pastoralism, the numbers of plants could decline sharply. The land is easily eroded by over-stocking and trampling by livestock is inimical to the development of seedling plants.

Farmers resident in the Estcourt area should be encouraged to contribute additional records of aloes that may be sighted outside the small known distribution range east of Estcourt. There is a good deal of suitable habitat in the Weenen Nature Reserve northeast of Estcourt. Should the species be found here, it would already be protected within the confines of a reserve. Various parts of the countryside near Estcourt are subject to land claims and it will be possible to assess the long-term future of *Aloe inconspicua* and other plants only once changes to the *status quo* have taken place.

Cultivation, propagation and horticultural potential

Aloe inconspicua is one of the most interesting grass aloes but is rather drab when compared to the other species in this group. It will probably appeal only to dedicated aloe collectors. The plants should be grown in fine shale gravel to which a small amount of sieved clayey soil has been added. This mixture ensures that the natural growing conditions are accurately imitated. The plants need a dry soil surface but some moisture content around the bulb, particularly at the base where the roots emerge. Clay containers are the most suitable for cultivating the plants, particularly those with a coarse finish permitting them to absorb excess soil moisture.

The aloes should be grown in direct morning sunlight and exposed to rain. They grow best where there is plenty of free air circulation. The bulbs should be watered in late spring, summer and early autumn, October to April in the southern hemisphere.

Seeds should be sown in the same mixture in which mature bulbs are cultivated. They are best sown on the surface of the soil in deep clay pots and kept in lightly dappled shade. Seeds should be just covered with soil. The pots should be regularly watered after sowing and throughout summer while the young plants are developing. Seeds germinate well when exposed to rain, but often rather slowly when tap water is used. They should be kept dry for the duration of autumn and winter. Watering should be resumed once more at the beginning of the following summer.

Aloe inconspicua flowers rather sparsely and erratically in cultivation and hand pollination is usually required to produce a good seed set.

11. *Aloe integra*

Aloe integra Reynolds

Aloe integra is found mostly in Mpumalanga and the adjacent parts of Swaziland. The grassland habitat where the plants occur is, in many cases, situated in regions with extensive timber plantations. In other areas the countryside has been invaded by introduced Australian trees such as black wattle (*Acacia mearnsii*). *A. integra* is locally plentiful but its habitat is limited and it is absent from several regions with hilly country that seem ideally suited to its requirements.

This species is found in much more arid habitat than all the other South African grass aloes of similar size. The plants grow mostly on north- and northwest-facing slopes with short grass where they are exposed to strong direct sunlight for the greater part of the day. *Aloe integra* is often found together with *A. maculata*, another species partial to hot, rocky slopes. In some years when *A. integra* flowers at the end of winter, its flowering period overlaps with that of *A. maculata*, resulting in occasional natural hybrids between the two species.

The aloes have been studied in two large populations that typify the current conditions under which the plants are found. One colony near Amsterdam in Mpumalanga occurs on a series of rocky hillsides facing north, about 2 km long. The other colony is situated on a number of large rocky hills mostly facing north and northwest near the resort town of Badplaas, also in Mpumalanga.

Near Amsterdam in Mpumalanga

The aloes in a large colony near Amsterdam, close to the border with Swaziland, occur along a range of steep, rocky hills above a river valley. Part of the habitat consists of rocky hillside grassland and the remainder comprises very stony slopes with sheets of exposed rock and tumbled boulders. The aloes are most common in the most rocky part of the habitat and occur mostly on the upper slopes of the hills. The grassy parts of the habitat are frequented by *Aloe kniphofioides*, which is quite often found beside *A. integra*. Plants of *A. integra* are absent from the south-facing hills on the opposite side of the valley. This shady hillside grassland is home to large numbers of *A. cooperi*.

Aloe integra flowers very well after veld fires have swept through the grassland in winter. The timing of flowering depends on the occurrence of the fires. If the grassland is burnt between May and July, the plants flower in August and September. Later grass fires result in a later flowering season, in September and October.

The habitat has been progressively degraded over the last 20 years. The south-facing hillsides and valley have been invaded by black wattles. These introduced Australian trees invade the moist parts of the habitat first and spread much more slowly to the drier, rocky, north-facing hillsides with shallow soil. Kikuyu grass is starting to

Right: ***Aloe integra*** is certainly one of the most handsome of all South African grass aloes, the bright lemon-yellow flowers held aloft above the lime-green leaves contrasting with the burnt grassland. These plants were painted on the mountains east of Badplaas in Mpumalanga in early August 2004.

transform parts of the grassland at the bottom of the hills and can be expected to spread further, particularly in heavily grazed sections of the habitat. Alien weeds proliferate in over-grazed areas, particularly *Verbena aristigera*. Indigenous annual plants such as the wild foxglove (*Ceratotheca triloba*) are also commonly found in large numbers on disturbed rocky soil.

There is good recruitment of young plants to the population in rocky areas with the shortest grass. Far fewer young plants are found in situations with the least rocks and dense short grass. *Aloe integra* grows sparsely in shallow soil on exposed sheets of rocks provided that there is some grass cover. Most of this type of habitat is densely covered by huge numbers of *Euphorbia pulvinata*. Some of these succulents form clumps almost 2 m wide. The aloes occur sparsely in a radius of some 50 m on top of the hills, mostly where the ground is rocky with short grass cover.

Plants of *Aloe integra* are very attractive to pollinators, particularly sunbirds (species of *Nectarinia*), when they flower in masses after winter grass fires. The racemes of bright yellow flowers are visible from a long way off. They furthermore appear at a time when there is little else by way of flowers in the surrounding veld, ensuring good pollination. This locality is one of the few places where *A. integra* still has its pollination biology intact and where fire cycles are not altered to protect timber plantations. This population is significant in that it provides a good indication of all the habitats in which the species occurs, as well as their paucity or abundance in each of these habitats.

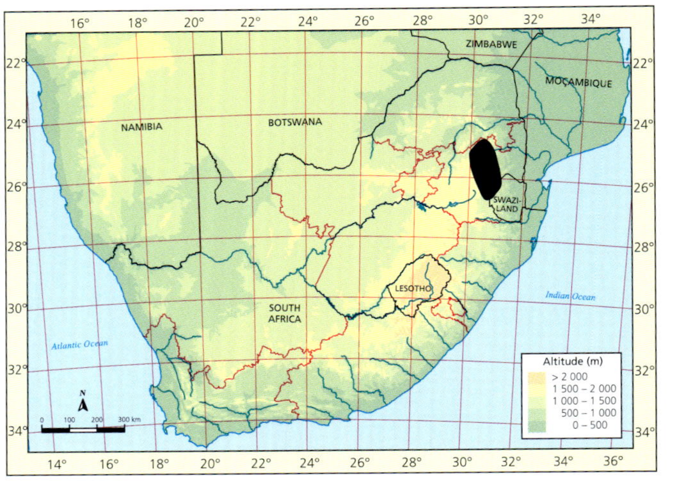

Distribution of *Aloe integra*

Near Badplaas in Mpumalanga

Plants of *Aloe integra* are found in a large, scattered colony on the mountains near Badplaas. Nearly the entire the habitat is forested with pines but the rockiest hills have escaped planting with timber. The unforested hills contain a significant amount of rocky grassland. The aloes occur mostly on the upper third of the slopes, except where the ground is very rocky in which case they are also found lower down. The plants inhabit mostly the north- and northwest-facing slopes, except for one section of the habitat where the aloes are also found on the rockiest topmost eastern slopes.

The veld is burnt nearly every year in May or June once the grass has dried at the beginning of the rainless winter. The burning is part of measures used to protect huge areas of highly flammable pine plantations. *Aloe integra* starts to form flower buds in late July and the first flowers appear in early August. The peak of the flowering season is usually the second half of August.

The seed set has been consistently limited for many years. The plants always flower well, but few of the main pollinators, such as sunbirds (species of *Nectarinia*), are attracted. Large areas of timber plantations exclude pollinators and this is one of the main problems facing *Aloe integra* with respect to its long-term survival. Recruitment of young plants to the population is negligible. If the limited amounts of seed do manage to germinate, the seedlings are destroyed in regular winter grass fires. The only places where young plants do manage to establish themselves are amongst rocks where they escape the hottest flames of the fires.

Two other problems affect the plants. Some of the young buds and shoots are frequently damaged by indigenous antelope at the end of winter. The buck do not graze the plants to any significant degree but are attracted by the bright green aloe leaves in the fire-blackened grassland. More recently baboons have started to use the lower sections of the flower stems as a source of food. The baboons snap off the stems and eat the lower portions, discarding the rest. During a field trip in August 2004 it was noticed that most flowers had been destroyed by baboons. The remainder of these stems were subsequently destroyed as soon as they lengthened, as they typically do just before the flower buds open. The damage by antelope and baboons occurs as a result of restricted foraging ranges, a consequence of extensive afforestation.

Growth cycle and the role of fire

Aloe integra is one of the earliest of the large grass aloe species to flower after winter grass fires. Flowering often occurs in August, the transition month between winter and spring. The plants initiate leaf development and bud formation in direct response to grass fires. The plants do not do well without the stimulation of fire. In places with thick grass cover flowering is minimal during the inter-fire years and usually stops altogether if the grassland is not burnt for three or four years in succession.

Mass flowering and seed production are restricted to seasons when there are winter grass fires. This ensures that seeds are liberated into the environment at a time when the conditions are the most favourable for germination. The aloe seedlings start their first growth cycle when they have least competition from grasses. The grass blades at this time serve the necessary purpose of filtering the sunlight on the hot north-facing slopes where *Aloe integra* normally occurs. The leaves of the aloes are very well adapted to burning. They form layers separated by a few millimetres, ensuring that the plants are thoroughly burnt during the course of a grass fire.

Right: ***Aloe integra*** is characterised by several layers of dead foliage in the dry winter months. The centre of the rosette of leaves, however, remains green. This plant was painted in early August, with day length increasing and ground temperatures beginning to warm up on the north-facing hill slopes frequented by these plants. Typical of the species, this plant has several almost acaulescent rosettes of leaves.

The aloes grow most actively just before, during and shortly after flowering. Seeds are liberated into the environment by wind and water runoff after rain. The period during which seeds are dispersed is particularly brief in this species. The seed capsules split open fully at seeding, an unusual phenomenon in grass aloes. Most species retain their seeds for at least a few weeks before the top of the capsule splits and the dispersal of seeds begins.

The aloes enter dormancy during April and May. The leaves have usually died back completely by the end of May, accelerated by the first frosts of winter.

Status of wild populations

It is important when evaluating the status of *Aloe integra* that a distinction is made between populations of which the biological processes that sustain them are intact and those of which the processes are impaired. There are few populations of this grass aloe that exist without substantial alterations to fire cycles or pollination biology. This will inevitably have significant implications for the long-term preservation of the plants.

Where *Aloe integra* grows in regions of afforestation, the grassland will have to be actively managed to secure the long-term survival of the plants. Efforts towards preserving the species are best directed towards populations of aloes such as the one near Amsterdam. Fire cycles and pollination biology are little altered here, but transformation of the grassland through invasive alien plants may become a chronic problem in the future. Landowners will play the most important role in the future preservation of *A. integra* since they are best placed to keep the veld free from the spread of aliens.

Cultivation, propagation and horticultural potential

Aloe integra grows well in cultivation, but rarely flowers. The plants should be grown in large, deep containers to accommodate the extensive root system. This species requires sharp drainage to grow well. It should be grown in a porous soil mixture consisting of equal parts coarse gravel or river sand, sieved sandy soil and sieved, very well-rotted compost.

The plants should be regularly watered throughout spring and summer. Watering should be reduced early in the autumn as the leaves start to die back and the aloes begin to enter dormancy. The aloes can be watered occasionally in the winter.

Aloe integra grows well in patches of short, rocky grassland that have been left intact in suburban gardens. This grassland can be burnt periodically in late winter to stimulate the aloes to come into flower. The plants can also be cultivated in gardens with steep slopes or embankments, ensuring sharp drainage. The aloes grow well in these situations but do not usually flower. To encourage flowering, dry veld grass should be heaped around the leaves and ignited in late winter. Once the dry leaves have been burnt, the plants should be thoroughly watered.

This species is readily propagated from seed. Seeds should be evenly scattered across the surface of the soil and very lightly covered as they are amongst the smallest of any grass aloe seeds. They are best sown in pots or deep seed trays, which should be well watered during the germination process and while the seedlings are developing. The young plants are prone to damping off, particularly if the soil is kept too wet and in situations where there is not enough free air circulation.

The seedlings should be grown in very light shade or in positions of morning sunlight. They should be kept dry during winter dormancy and benefit from an occasional light watering about once every six weeks in winter. The young plants are best transplanted at the end of their second or third growing season, once they have a few well developed leaves.

Aloe integra is one of South Africa's most spectacular flowering plants. It is, however, unlikely to become popular in horticulture as it is more difficult to cultivate than the other large grass aloe species.

12. *Aloe kniphofioides*

Aloe kniphofioides Baker

This is one of many grass aloe species that have been negatively affected by the advent of timber plantations in South Africa. Plants of *Aloe kniphofioides* used to be common and widespread in mist belt grassland near Piet Retief in Mpumalanga. The species was also well represented in adjacent regions to the north near Amsterdam and The Brook. These regions have been extensively afforested with eucalypts and pines, with the result that most of the former habitat has been destroyed.

The species used to occur in large numbers south and southwest of Piet Retief. This area has been extensively degraded by invasive alien trees, mostly black wattles (*Acacia mearnsii*). More recently, habitat not invaded by wattles has been converted to open-cast coal mines, particularly in the vicinity of the small town of Dirkiesdorp.

The aloes used to be frequent north of Lochiel in northeastern Mpumalanga. This region has now been afforested where it lies within the mist belt. Further north where the hilly country is lower the habitat has been degraded by over-grazing and too frequent burning.

The species has fared little better in KwaZulu-Natal, which was once its second stronghold after Mpumalanga. Plants in the Vryheid District have all but disappeared owing to afforestation. The places where eucalypts and pines grow best in Mpumalanga and KwaZulu-Natal are also the optimum habitat for *Aloe kniphofioides*. Its pollination biology has been seriously affected by timber plantations. In former times, sunbirds (species of *Nectarinia*) were the main pollinators. These active nectar-seeking birds were attracted by large areas of grassland full of the bright orange flowers of *A. kniphofioides*, the inflorescences conspicuous from afar in the sere blackened grassland after winter veld fires. The species now occurs on small strips of grassland surrounded by timber plantations, as a road reserve plant and, more rarely, on large rocky hillsides unsuited to eucalypt and pine plantations.

As there are still a number of pockets of grassland with *Aloe kniphofioides*, the impression is created that the aloes are still a significant component of the remaining grassland habitat. Most plants are, however, old specimens that are remnants of much larger populations before afforestation.

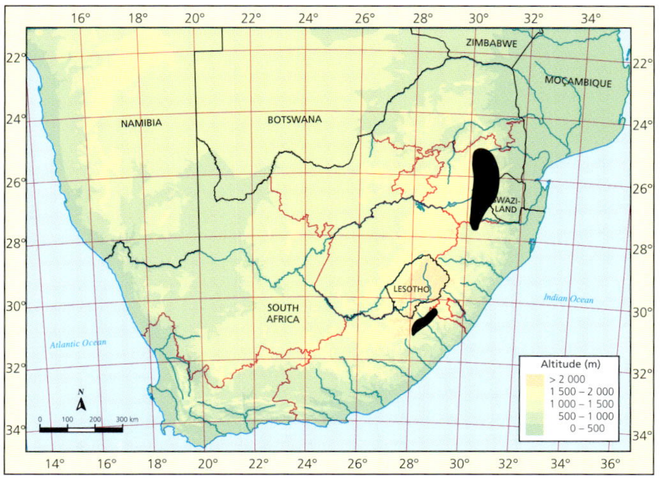

Distribution of *Aloe kniphofioides*

There is a negligible recruitment of young plants to most of these populations. This is related to the lack of pollinators and the annual burning of the strips of grassland as firebreaks for the protection of timber plantations.

Over the last decade, some eucalypt and pine plantations near Piet Retief have been felled as the timber has reached maturity. The debris from the felling operations has been ignited and the burning has brought *Aloe kniphofioides* into flower. It is remarkable that these plants have survived for about 20 years in the thick plantation in their transformed habitat, covered by dead eucalypt leaves or pine needles. These areas are usually replanted with a second crop of timber within a short space of time.

Aloe kniphofioides has been intensively studied throughout its distribution range in KwaZulu-Natal and Mpumalanga over a period of some 20 years. The account that follows concerns various populations in their present environment.

Near Piet Retief

Already in the 1980s *Aloe kniphofioides* was becoming rare near Piet Retief. There were a few large populations, mostly in strips of grassland along watercourses. Some of this habitat stretched for many kilometres and included representative transects of the grassland habitat before afforestation. These colonies sometimes numbered several thousand plants and were sufficiently attractive to pollinators to ensure a satisfactory seed set. The veld was usually burnt each winter as part of fire prevention measures for areas near the plantations.

The regular fires ensured that the plants flowered in October and early November each year. There was enough seed produced to ensure that some regeneration took place. Annual fires are not, however, ideal for this species, restricting seedlings to rocky ground close to large tufts of grass. In these positions the young aloes were protected from the hottest flames.

Over the last 15 years the plantation belts were extended right down to the banks of the watercourses in some cases. The only ground left intact was that which was too rocky for trees to be planted. At this time the numbers of *Aloe kniphofioides* were too low and the populations too fragmented to attract pollinators when the plants were in flower.

In the Piet Retief District *Aloe kniphofioides* is partial to areas of deep soil with short grass. Most of this habitat is now under timber plantations and the largest amounts of terrain that remain are characterised by rocky ground. The species does occur in these situations throughout its distribution range, but the plants are rarely plentiful.

The aloes growing along river or stream corridors of grassland set seed rather erratically. The larger and wider these corridors, the greater the likelihood that there will be enough flowers to attract pollinators. Groups of plants growing in narrow strips of grassland surrounded by plantations rarely set any seed from one year to the next. The seed that is set is too limited to maintain populations of this species in these areas in the long term.

In the Dirkiesdorp area of Mpumalanga

Aloe kniphofioides used to occur in the hilly grasslands around Dirkiesdorp, its habitat consisting of gently undulating ground near streams and rocky hillsides. In recent decades some of the habitat has been destroyed by open-cast coal mines. The areas around the streams have been subject to invasion by black wattles for many years and much of the habitat has been transformed. The wattles have now started to become frequent in areas of more open veld. Some of the area around Dirkiesdorp has also been planted with timber but to a much less extensive degree than near Piet Retief. *A. kniphofioides* is currently found on steeply sloping hills with short grassland.

Near Amsterdam in Mpumalanga

The grasslands around Amsterdam have largely been destroyed by eucalypt and pine plantations. The remaining habitat consists of rocky hills unsuited to the establishment of plantations. One of the largest remaining colonies of *Aloe kniphofioides* is found on the slopes of an extensive series of rocky hills southeast of Amsterdam. The plants are most frequent in patches of deep soil near stream banks and in depressions on the hillsides which also hold deep soil.

The habitat is burnt every few years once the grass cover is sufficiently dense. Flowering is prolific after fire but the seed set has been poor in recent years. This situation is not readily explained as the area to the east contains similar habitat without timber plantations. The kinds of factors requiring further research are the range of pollinators in the area when the aloes are in flower and whether or not other sources of nectar vie for the attention of prospective pollinators. It has, however, become evident that the numbers of sunbirds (species of *Nectarinia*) in any district are reduced through extensive monocultures of eucalypts and pines.

Near Amsterdam *Aloe kniphofioides* occurs on level ground along a river. The plants also frequent the steep north- and west-facing hill slopes where they grow together with *A. integra*. *A. kniphofioides* is absent from the south- and southwest-facing hill slopes on the opposite side of the river. The plants are rather sparse and widely scattered on the stony hillsides but more plentiful on the level ground. In recent years the flatter areas have been heavily trampled by grazing livestock, which has led to a decline in the number of aloes. This is one of the few remaining localities where the range of habitats available to *A. kniphofioides* have not been afforested with timber.

Near Lochiel in eastern Mpumalanga

The Lochiel area is another region densely afforested with eucalypts and pines. *Aloe kniphofioides* has been all but eradicated from this region. North of Lochiel the plants are found at the edge of the plantation belt, in rocky ground. This area has been degraded by over-grazing and too frequent burning as well as invasion by black wattles. The aloes are thinly distributed over stony hillsides in areas where the soil is deep enough.

Seeds are set but rather sparingly owing to the degraded habitat, which does not have enough plants in flower to attract pollinators. Regeneration of *Aloe kniphofioides* in rocky habitats such as this is also particularly slow compared to regions of short grassland with deep soil.

The aloes are also found sparsely further north. The hills to the north of the timber plantation belt are at a lower altitude than the Lochiel area and lie outside the mist belt. *Aloe kniphofioides* is restricted mostly to the southern and eastern slopes of grazing hills in the vicinity of seepage areas or near drainage lines. In recent years the area has been settled by rural pastoralists and the veld has become degraded and over-grazed. *A. kniphofioides* populations in this area consist mostly of mature plants that are better able to withstand trampling by livestock.

Near Vryheid in KwaZulu-Natal

Much of the grassland in the mist belt near Vryheid has been forested with eucalypt plantations. *Aloe kniphofioides* has become very rare in the area since the extensive development of timber plantations. The aloes are found mostly on rocky hillsides, the least suitable habitat and a repetition of the circumstances prevailing in Mpumalanga.

The hilly grassland habitat that remains is wedged in between plantations and coal mine dumps. These remnants of formerly extensive grassland provide the only suitable grazing for livestock in heavily afforested areas. Trampling restricts the plants to the rockiest areas of the habitat, which contain the shallowest soil. Regeneration from seeds is very limited, restricted to the few seedlings that manage to survive in rocky ground.

Growth cycle and the role of fire

Aloe kniphofioides starts its growth cycle early if the veld has been burnt the previous winter. New leaves start to emerge in the second half of August and flower buds start to form in late August or early September. In grassland that has not been burnt the leaves sprout much later, usually at the end of September or early October. The flowering season is mostly October in areas where there have been fires the previous winter. In grassland that has escaped burning, fewer plants flower and the flowering season is usually in November.

Fire is essential for the life cycle of *Aloe kniphofioides*. The plants flower in mass after a winter grass fire, attracting the greatest number of pollinators. The seeds are furthermore liberated into the environment at a time when conditions are most conducive to germination and seedling development. There is little competition from dense grasses and all the places frequented by this species are opened up after a fire. Flowering decreases each successive year without burning, only for mass flowering to resume again after the next fire.

Annual burning, which is a regular practice in timber-producing areas, stimulates flowering each year. There is however, too little opportunity for seedling development where the veld is burnt annually. Under such circumstances, regeneration is restricted to parts of the habitat where the young plants escape the hottest flames—the rocky areas where there are fewest plants owing to the shallow soil.

The aloes set seed in late December and early January. The seeds are dispersed by wind and water runoff after rain. Some seeds germinate soon after they have been shed and the seedlings have some eight to 12 weeks to develop before the beginning of cooler, dry autumn weather.

Right: ***Aloe kniphofioides*** growing in a rocky road reserve near Piet Retief in Mpumalanga. The veld had been burnt the previous winter, a prerequisite for mass flowering in this species in spring and early summer.

Status of wild populations

Aloe kniphofioides has declined to a fraction of its former numbers and continues to decline as more of its habitat is degraded and destroyed. The long-term survival of the plants is precarious in a landscape dominated by timber plantations and open-cast coal mining. In many areas, *A. kniphofioides* can be expected to die out during the course of this century.

Efforts to preserve these aloes are best directed towards conserving the largest populations of the species, where pollinators still visit the flowers and there is a satisfactory seed set. *Ex situ* conservation may be an option. This subject would merit debate amongst various specialists during the next few decades.

Cultivation, propagation and horticultural potential

With its upright leaves and racemes of striking orange or reddish orange flowers, *Aloe kniphofioides* is a handsome plant, but it is difficult to bring into flower regularly in cultivation.

The aloes should be grown in a sharply drained soil. They thrive in a mixture of one third sandy soil, one third river sand or fine gravel and one third very well-rotted compost. finely sieved peat moss can be used instead of compost. The aloes grow best in earthenware containers as they absorb any excess moisture. They should be grown in positions of morning sunlight or lightly dappled shade. In summer-rainfall areas, the plants can also be cultivated in a rockery provided that the position has deep soil and is very well drained. In late July or early August dry veld grass should be heaped around the dead leaves of adult plants and ignited. The aloes should be regularly watered afterwards. The burning usually stimulates flower bud formation a few weeks later.

The plants should be well watered throughout spring and summer. Watering should be reduced in autumn and the aloes should be kept completely dry for the duration of the winter.

Aloe kniphofioides rarely sets any seed in cultivation without meticulous and regular hand pollination while the plants are in flower. Seeds should be sown in late spring. It is best to scatter them over the surface of the soil and cover them lightly. Deep seed trays or earthenware pots are the most suitable. It is important to ensure that a good layer of coarse drainage is packed at the bottom of the containers used for propagation. Seeds can be sown in the same mixture used for the cultivation of adult plants. The soil should be kept moist throughout the germination period and for the remainder of the summer while the young plants are developing. They should be kept dry and in a cool, shaded place for the duration of winter. Seedlings of *A. kniphofioides* are usually ready for transplanting at the beginning of their third growing season.

Aloe kniphofioides starting its growth cycle in early August after a winter grass fire near Piet Retief in Mpumalanga. The burnt veld allows the ground to warm up, stimulating the aloes to sprout new leaves. The painting also depicts two bulbs. One is that of a mature, well-established plant with its extensive root system. The other bulb is estimated to be some three years old. The growth rate of seedlings of this species is much slower in the wild than in cultivation.

13. *Aloe kraussii*

Aloe kraussii Baker

Aloe kraussii is a striking yellow-flowered grass aloe that has become increasingly rare owing to the afforestation of large areas of its former habitat with timber plantations. In other areas where the plants occur, the veld has been degraded with invasive alien vegetation, particularly black wattles (*Acacia mearnsii*) and silver wattles (*A. dealbata*). Most of its former habitat has been urbanised or built up with informal settlements in areas between Pietermaritzburg and Durban in KwaZulu-Natal where the species used to occur.

The aloes used to be well represented in the Greytown District of KwaZulu-Natal. This area is now so heavily afforested with aliens that only small pockets of grassland habitat that support the aloes, remain.

Reynolds (1969: 145) considered *Aloe kraussii* to be allied to *A. ecklonis*. Glen & Hardy (2000: 19) reduced *A. kraussii* to synonymy under *A. ecklonis* but gave no reasons for their decision. This is a problem with respect to advancing knowledge about South Africa's grass aloes. It will be necessary for detailed sampling of both species in areas where they grow together to establish the validity or otherwise of this reduction. *A. kraussii* grows alongside *A. boylei* near Greytown and the flowering season of the two aloes, in the second half of December, usually overlaps. Both species are pollinated simultaneously by black sunbirds (*Nectarinia amethystine*) and very limited hybridisation occurs. This is the kind of information that population studies reveal and usually proves the most useful for assessing the distinctiveness of described grass aloe species during the course of taxonomic revisions. Habitats of *A. kraussii* have become very fragmented in recent decades and it is therefore unlikely that there are many opportunities for studying this species where it grows together with *A. ecklonis*.

Aloe kraussii is partial to mist belt habitats. It also occurs at the edge of mist belts but in small numbers, such as the hilly country to the west of the Mooi River Falls in KwaZulu-Natal.

Reynolds (1969: 144) refers to aloes similar to *Aloe kraussii* that were found in the mountains south of Barberton in Mpumalanga and thought they were likely to be *A. ecklonis*. These plants appear to be the recently described *A. craibii*, a species which is superficially similar to *A. kraussii*.

Aloe kraussii has been studied for several years in the Hermannsburg area where it grows in a small grassland reserve belonging to the Hermannsburg School Trust and is monitored by the Deutsche Schule in Hermannsburg. The area is densely afforested with black wattles, eucalypts and pines. There are, however, a number of significant riverine grassland corridors that link up with more elevated hilly grassland nearby.

Near Hermannsburg in the Umvoti District of KwaZulu-Natal

Some years ago, a few dedicated naturalists living in Hermannsburg made efforts to preserve a strip of grassland particularly rich in the local indigenous flora. The area consists of a west- and southwest-facing hill slope with a stream running below. This area is only a few hectares in extent and is hemmed in by timber plantations on all sides, except for the veld near the stream and its tributaries. The

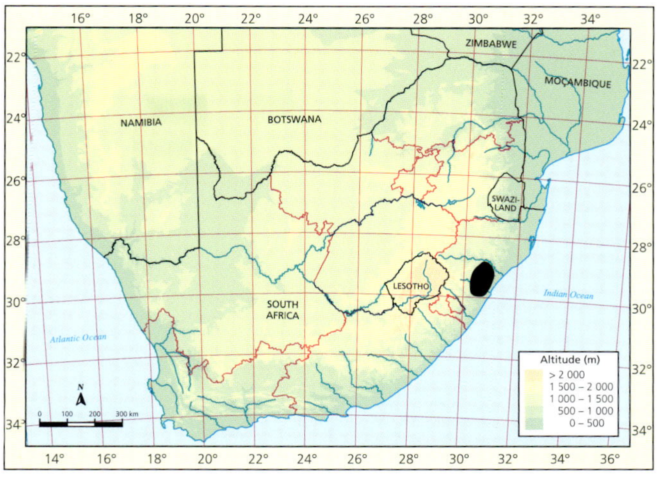

Distribution of *Aloe kraussii*

grassland is fenced off and for some years it has not been subject to the usual grazing pressures from livestock associated with the strips of land between timber plantations in the area.

The veld is burnt periodically in winter, leading to mass flowering of the aloes. *Aloe kraussii* occurs on the upper grassy slopes whereas *A. boylei* is found growing lower down amongst rocks close to the river. The two species are found within about 30 m of one another.

Aloe kraussii has increased in numbers in this stretch of conserved grassland. The aloes occur in clumps of individual plants or singly, scattered across the veld. Most plants grow in deep soil with dense but short grass cover. They also frequent the edges of rock sheets in the fenced grassland and the adjacent road reserve. A number of other plants sensitive to trampling by livestock are found, including *Drimia uniflora*, a tiny bulb that grows in thin soil cover and moss.

An extraordinary attribute of this habitat is the fact that *Aloe kraussii* is very well pollinated whenever flowers are produced. Mass flowering usually follows a grass fire the previous winter. The large numbers of yellow flowers are very attractive to honeybees (*Apis mellifera*) and black sunbirds (*Nectarinia amethystine*). Sunbirds are usually absent from areas of extensive timber plantations. Their presence is attributable to significant stretches of grassland along rivers and streams.

The grassland reserve near Hermannsburg provides a significant example of how veld can be preserved with fire cycles and pollination processes intact. This stretch of grassland can be used as a model for the preservation of similar areas in other regions heavily afforested with aliens.

Aloe kraussii also occurs as a road reserve plant close to Hermannsburg. The aloes grow in narrow strips of grassland bordering timber plantations. They are not numerous owing to the limited habitat available to them, but are visited by pollinators and set some seeds when flowering is good after grass fires the previous winter.

Growth cycle and the role of fire

This summer-growing species sprouts leaves with the beginning of the main summer rains in late October and early November. Rapid leaf growth is followed by flower bud formation in early December. The peak of the flowering season is late December and early January.

The plants are in seed in February. Seeds are dispersed

by wind and water runoff after rain. The seeds are small and light, easily blown about in the wind. February is often a misty, rainy month where *Aloe kraussii* occurs. The seed capsules require strong, direct sunlight to split open fully. In misty or rainy weather this process is delayed and seeds are retained in their capsules until well into March or early April.

Winter grass fires create the ideal conditions for *Aloe kraussii* to flower the following summer. They clear the veld from moribund grasses and open up the habitat for seed dispersal. Seedlings of this species develop rapidly and are usually able to withstand winter fires at the end of their second growing season. Where seeding is prolific, such as in the grassland reserve near Hermannsburg, there are enough young seedlings established in different places to allow some to survive consecutive years of grass fires.

Aloe kraussii start to die back in late April. The foliage has usually withered by mid-May. The prostrate dead aloe leaves form a thick mat that ensures that the stems of the plants are thoroughly burnt during the course of a grass fire.

Status of wild populations

Aloe kraussii is one of the rarer grass aloes and is unlikely to survive in the long term unless efforts are made to preserve it in grassland where its biological processes remain intact. The plants are now extinct in a number of the habitats recorded by Reynolds (1969: 144).

If the view is taken that *Aloe kraussii* was correctly reduced to synonymy under *A. ecklonis*, then it is little more than a regional variant of *A. ecklonis*, the most common and most widespread grass aloe species in South Africa. Alternatively, if *A. kraussii* is upheld as a species, it takes its place amongst the rarest South African grass aloes existing only in some parts of its former habitat. The perspective adopted here is that the status of *A. kraussii* has to await detailed sampling of different populations as well as innovations in taxonomic theory and model construction. At this stage of the knowledge construction process it seems logical to analyse the species concept of *A. kraussii* with reference to its nearest affinities.

Cultivation, propagation and horticultural potential

This species is easily cultivated in a well-drained soil. It grows well in a mixture consisting of half gravelly sand and half sieved, well-rotted, acid compost or peat moss. The plants have a deep root system which is best accommodated in large, preferably earthenware containers. These containers absorb excess moisture, but plastic pots are also suitable if provision is made for adequate drainage.

Aloe kraussii can be grown either as a container plant or a rockery subject. The aloes require a deep, sandy, well-drained area when they are grown in a rockery. Plants grown in a rockery thrive in day-long sunlight while those in pots dry out too quickly if they receive sunlight all day. Aloes in containers are best grown in positions of morning sunlight.

The plants should be regularly watered throughout the summer growing season. They grow best when exposed to rain. They should be kept dry during the winter months except for an occasional watering during mild weather in late winter.

Aloe kraussii is readily propagated from stem cuttings and seeds. Cuttings are difficult to take since this species has very short stems. A sharp knife should be used for taking cuttings and they should be removed as low down the stem as possible without damaging the stem material of the parent plant. Cuttings should be taken in late winter and early spring and rooted in moist, coarse gravel or river sand. A root system has usually developed by late spring or early summer and the stems are ready to transplant at this stage. Cuttings can also be struck in autumn as night temperatures drop. They root rapidly at this time but should be transplanted only the following spring. They should be lightly watered at intervals during winter to protect the new root growth against desiccation.

Some seeds are produced in cultivation with the aid of hand pollination. The seeds should be evenly scattered across the surface of the germination medium in spring or early summer and just covered with soil. The mixture used for cultivating mature plants in containers is also suitable as a germination mix. Seeds should be sown in deep seed trays or pots which have to be kept moist throughout the germination period and afterwards for the duration of the summer while the young plants are developing. Containers used for germination are best kept in positions of morning sunlight with good ventilation. Seedlings are ready for transplanting at the beginning of their third growing season.

Aloe kraussii is a showy species on account of its striking yellow flowers and bright green leaves. These aloes should become as popular as *A. cooperi* in cultivation if they are readily available in nurseries.

Right: ***Aloe kraussii*** in a privately managed grassland reserve near Hermannsburg east of Greytown in KwaZulu-Natal. Seedlings such as those in the foreground are regularly encountered in the reserve.

J Condy

14. *Aloe linearifolia*

Aloe linearifolia A.Berger

This is another grass aloe that has had much of its former habitat destroyed. Most grasslands where the plants used to be widespread along the south coast of KwaZulu-Natal have now been converted to eucalypt plantations and sugar cane fields. Reynolds (1969) also recorded the species from south-central KwaZulu-Natal and eastern the sections of the Transkei. There is little modern information on the species in these areas which have both witnessed a large expansion in the rural population accompanied by overgrazing and the degradation of grassland habitats.

Aloe linearifolia is sometimes found in small groups at the edge of eucalypt plantations, usually growing on rocky ground unsuitable for planting timber. These aloes are the remnants of much larger colonies of this species which used to be frequent in the undulating coastal grassland. Some plants still occur within eucalypt plantations, producing a few leaves but rarely any flowers.

The species may never have been widespread in the coastal grasslands of KwaZulu-Natal. Prior to the extensive development of agriculture and urbanisation on the south coast, *Aloe linearifolia* was regarded as rare (Reynolds 1942). Subsequent accounts of the species have not, however, contained specific comments as to its rarity or distribution in grassland. *A. linearifolia* has been studied at three localities in KwaZulu-Natal, all of which indicate the modern environmental pressures encountered by this species. The aloes have been observed for some years in the hilly coastal grassland on the northern outskirts of Margate, near Fair Acres at the eastern end of the Oribi Gorge and in the grassland of the Kranskloof Nature Reserve within suburban Durban.

In the Kranskloof Nature Reserve

Aloe linearifolia is still numerous in the Kranskloof Nature Reserve, but its habitat has been reduced by suburban development on either side of the reserve. The remaining habitat consists of short stretches of grassland suburbia and the vast cliff faces flanking the river gorge in which the reserve is situated. The habitat consists of gently undulating hills above the gorge intersected by steep ravines harbouring dense indigenous bush.

The aloes are widely scattered across east-, south- and west-facing hill slopes. Plants growing in deep soil are robust, with several branches, while those growing in rocky ground or in places between boulders, produce only a few stems. The colonies contain plants of various sizes ranging from mature aloes down to seedlings, indicative of healthy regeneration. Although the plants are relatively numerous, they occur only in a fragment of the hillside territory that would once have been available to them before the development of suburbia.

The plants flower mostly in early February. There is usually a good seed set, indicating that suburban development has not interrupted the pollination biology of this species. The grassland was in good condition at the time the data for this book were collected. There were, however, signs of infestation by alien invasive plants, particularly bugweed (*Solanum mauritianum*).

The Kranskloof Nature Reserve may be one of the few places where *Aloe linearifolia* is conserved in significant numbers in the future.

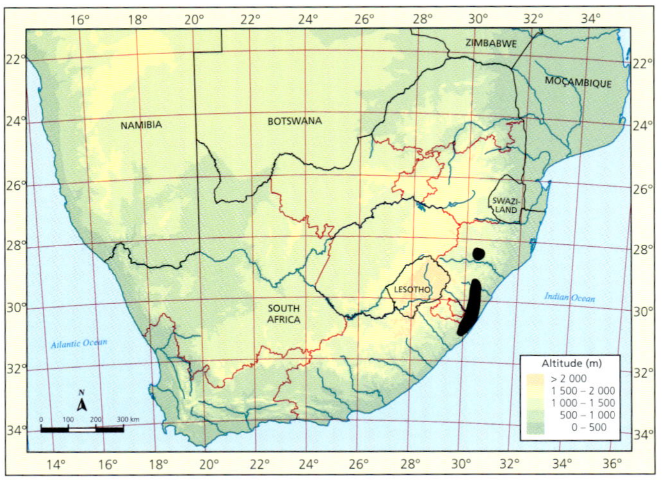

Distribution of *Aloe linearifolia*

On the northern outskirts of Margate

The hilly grassland north of Margate has significant colonies of *Aloe linearifolia*, but they are decreasing with the passage of time as more and more of the coastal grassland is taken over by suburban developments. The area inhabited by the aloes lies between suburbia on the seaward side and large monocultures of sugar cane fields and eucalypt plantations on the landward side.

The plants have been studied for about 15 years and this has provided a good opportunity to assess their distribution patterns in hilly coastal grassland. Initially the flatter areas, which supported good numbers of plants, were undeveloped. Suburban developments have taken place here over the last few years, eliminating the aloes from the lower lying areas.

Aloe linearifolia has an interesting distribution in hilly grassland. It is found on all aspects of the hills, including the hot northern slopes which are in direct sunlight for much of the day. The grass cover is uniformly short across the habitat, much of which consists of rocky outcrops, particularly on the summits of the hills. The lower lying places along drainage lines are truncated by patches of natural bush, with large numbers of *Strelitzia nicolai*. The plants used to be at their most numerous in flattish areas at the base of the hills where the deep soil remains moist for long periods after rain. This area has now been developed, but a few aloes still remain on the suburban properties of people who have left the grassland intact.

The remaining habitat has two principal habitats where the plants are found—the grassland in the vicinity of rocky outcrops and steep hillsides that slope away towards the drainage lines. Occasional plants are found in shallow sandy depressions on sheets of exposed rock.

The habitat is undergoing degradation through invasion by alien weeds and illicit dumping of building rubble and garden refuse. Garden refuse often contains kikuyu grass (*Pennisetum clandestinum*) rhizomes, which begin to transform the grassland. It is a pernicious transformer of grassland and difficult to eradicate once it has taken root.

Right: ***Aloe linearifolia*** growing in gently undulating hilly grassland on the outskirts of Margate in KwaZulu-Natal. The plants are usually most frequent in well-drained, stony soil amongst short grass, often on rocky outcrops.

As a result of the unauthorised dumping of refuse, various garden plants with invasive tendencies are becoming a problem, degrading and transforming the grassland. These include cannas (*Canna indica*) and Madagascar periwinkles (*Catharanthus roseus*).

Near Fair Acres

Plants of *Aloe linearifolia* are sparsely found in the veld near Fair Acres where their habitat comprises the remains of a much more extensive area of short grassland. It is hemmed in by sugar cane fields and invasive black wattles (*Acacia mearnsii*) on one side and bordered by the cliffs at the eastern end of the Oribi Gorge on the other.

Most of the *Aloe linearifolia* specimens are large, with several thick stems. Smaller plants are more rarely found, usually growing in sandy soil at the edge of sheets of exposed sandstone. This may be related to the past infrequency of fires and the relatively thick grass cover, circumstances unsuited to the development of grass aloe seedlings. The remaining areas of relatively pristine grassland habitat were in the initial stages of invasion by black wattles while the data for this book were being collected.

Growth cycle and the role of fire

Aloe linearifolia is a summer-growing species producing foliage in November and December. The plants start to form flower buds in the second half of January. The peak of the flowering season is the first two weeks of February, but occasional plants may flower until late March. Seeds are liberated into the environment during the last few days of March and the first half of April.

Seeds are distributed by frequent breezes and runoff after rain. In places where *Aloe linearifolia* is still well represented, plants sometimes grow in scattered groups consisting of mature plants as well as young ones of various sizes. This is an indication that seeds often germinate in the vicinity of the parent plants.

Fires clear the habitat periodically of dead grass. This is necessary for the health of the aloe stems and allow enough sunlight to reach developing seedlings and juvenile plants. The aloes flower consistently during inter-fire years until the veld carries a heavy load of dead grass, usually about three years after the last grass fire. At this time, flowering performance tapers off until the next burn.

Status of wild populations

There is little recent information on the occurrence and distribution of plants in the Lusikisiki District of Transkei. The same applies to the Babanango area in Zululand. These were districts in which *Aloe linearifolia* was once recorded (Reynolds 1969: 141). The grassland of the Transkei is now densely and heavily grazed and the area near Babanango has been densely afforested with eucalypts and pines.

The aloes continue to decline near Margate as a result of urbanisation and degradation of the habitat with invasive alien plants. Most of the grassland along the KwaZulu-Natal south coast has been converted to a monoculture of sugar cane or is densely settled. Most of the remaining habitat is afforested with timber plantations.

Few remaining large populations of this grass aloe are likely to remain intact in the long term. Only one area with substantial numbers of plants is afforded formal protection within the confines of the Kranskloof Nature Reserve.

Urban developments have not altered the pollination biology of the species. This is a significant factor that has to be taken into account for the preservation of *Aloe linearifolia* in the future.

Cultivation, propagation and horticultural potential

Aloe linearifolia is best cultivated in large, preferably earthenware containers or well drained, undisturbed garden beds rich in sandy soil. A good cultivation mixture for growing plants in containers and for sowing seeds and raising seedlings, consists of half sieved sandy soil mixed with equal parts fine silt, coarse gravel and very well-rotted, sieved leaf mould or veld grass compost.

The aloes should be well watered throughout the late spring, summer and early autumn. Watering should be gradually reduced as the leaves start to die back and the plants enter dormancy in the autumn. *Aloe linearifolia* requires plenty of free air circulation while it is in leaf and also while flowering. Plants grown in confined spaces are prone to attack by root mealybug and also do not flower very well.

Seeds should be lightly and evenly scattered across the surface of the soil, and just covered. They should be well watered until germination and afterwards while the young seedlings are developing. It is important to keep the propagation trays well ventilated since *Aloe linearifolia* seedlings, like those of other grass aloe species, are prone to attack by damping-off fungi. The seeds are best sown in pots with a good soil depth. The species has long, thickened roots that require plenty of room to develop. Seedlings should be transplanted in their third growing season. Seed set is usually poor in cultivation, without hand pollination.

Plants can also be grown from cuttings or suckers struck in coarse sand that is kept moist while they are forming roots. Cuttings strike most readily in early summer or early autumn but are prone to rot if taken during the hottest part of the summer.

15. *Aloe micracantha*

Aloe micracantha Haw.

This is one of the more robust grass aloe species. It inhabits coastal grassland in the Eastern Cape and is also found inland in hilly and mountainous areas around Grahamstown. The distribution extends from the Humansdorp and Joubertina areas to the elevated grasslands east of Grahamstown. It may occur in Ciskei east of the Fish River but, if so, populations are likely to be small and heavily overgrazed. Some of the colonies of plants on the outskirts of Port Elizabeth are rather large and occur in what is still relatively pristine coastal grassland. As the suburbs of Port Elizabeth expand, these habitats will be destroyed.

The dumping of domestic and garden waste and extensive invasion by alien vegetation is a serious threat to many of the remaining colonies of plants south and west of Port Elizabeth. Garden refuse often contains alien plants such as rhizomes of kikuyu grass (*Pennisetum clandestinum*) which invade the surrounding veld.

The southern part of the distribution range lies at the transition between winter and summer rainfall. Throughout the distribution area, but particularly in the south, rain falls at any time of the year, with a tendency towards spring and autumn. Winters are on average relatively dry and plants are mostly deciduous and dormant during the cooler winter months.

Plants in the coastal grassland around Port Elizabeth have thin canaliculate leaves, whereas in two colonies near Grahamstown the leaves are much broader and lanceolate. It was at first thought that plants in these populations may be natural hybrids with *Aloe lineata* but this is unlikely. The flowers conform in every respect to the description for *A. micracantha* and the leaves die back in the winter, typical of all grass aloe species.

David Cumming of Bathurst and Charl Malan of Grahamstown kindly supplied valuable information about plants in habitat near Grahamstown. Welland Cowley of Port Elizabeth provided substantial and useful information about the species in coastal grassland to the southwest of Port Elizabeth. The account that follows focuses on the aloes in these two parts of their distribution range.

Near Greenhills east of Grahamstown

The area between Grahamstown and the farm Greenhills to the east has been extensively invaded by alien vegetation, particularly black wattles (*Acacia mearnsii*), pines and also the prolific Port Jackson willow (*A. saligna*). *Aloe micracantha* may well have been well represented in this area before extensive invasion by alien plants. It is now confined to steeply sloping hills densely covered with tall grass and low scrub. The typical habitat is illustrated in the pencil drawing on page 11.

The hillside where the aloes occur consists of steeply sloping, dense grassland interspersed with Suurberg cushion bush (*Oldenburgia grandis*), with little habitat for aloes that have to compete with tall grasses. Consequently the population is very small, consisting of a handful of mature plants, which had nearly finished flowering when the site was visited on 17 February 2004. Very little seed had been set in this population, probably as it is too isolated in dense grassland to attract many pollinators.

The steep habitat on the lower slopes supports a large number of *Aloe lineata*, which were starting to come into flower at the time of the visit. The northeast-facing slopes on the opposite side of the valley are heavily infested with black wattles and it is likely that this small population of *A. micracantha* would also be smothered under wattles in due course. The habitat at the Greenhills site is typical of areas near Grahamstown. The countryside has too many shrubs and tall grass to permit the development of sizeable populations of *A. micracantha*, such as the large colonies encountered south and southwest of Port Elizabeth.

In the Featherstone Kloof

The Featherstone Kloof has been cleared of extensive tracts of alien trees in recent times. The habitat restoration project concerned is discussed in detail under *Aloe myriacantha*. *A. micracantha* is found in limited numbers on the hills above the Featherstone Kloof and on the lower southwest-facing convoluted hill slopes that have been cleared of dense thickets of alien vegetation. It is now found largely on the rocky slopes amongst grass in habitat that was only sparsely occupied by alien vegetation before the veld restoration project.

There are many similarities between the Greenhills site and the habitat encountered at Featherstone Kloof. The hills are steep and rocky, with *Aloe micracantha* largely confined to the central and upper slopes and summits of hills. *A. lineata* is common on the lower third of the southwest-facing slopes and grows together with *A. micracantha* at some places. The Featherstone Kloof was visited on 16 and 18 February 2004. The *A. micracantha* had largely finished flowering and the seed set was poor, except for one plant growing in an exposed position on a rocky outcrop.

It is probable that the small isolated groups of *Aloe micracantha* in the Featherstone Kloof are not sufficiently conspicuous to attract pollinators in most flowering seasons. This is in marked contrast to *A. lineata* which flowers abundantly on the slopes of the Featherstone Kloof. These conspicuous flowers attract insects and sunbirds and are consequently easily able to produce large amounts of seed. A study of the relationship between *A. micracantha* and its pollinators in the habitat around Grahamstown would be useful. It would provide information as to whether or not these plants could re-establish themselves via a new seed bank on land cleared of aliens.

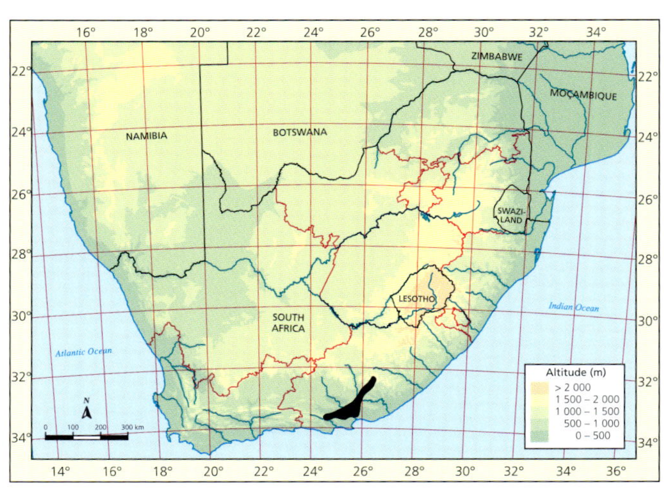

Distribution of *Aloe micracantha*

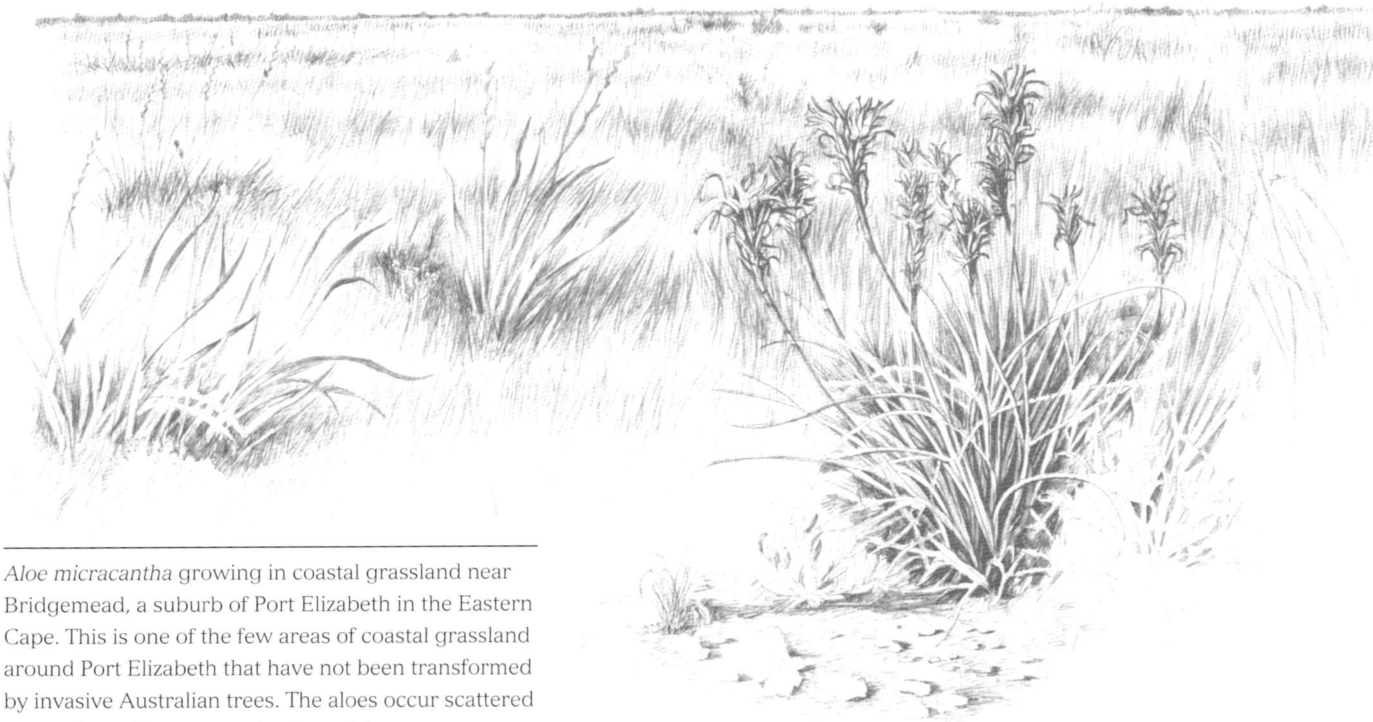

Aloe micracantha growing in coastal grassland near Bridgemead, a suburb of Port Elizabeth in the Eastern Cape. This is one of the few areas of coastal grassland around Port Elizabeth that have not been transformed by invasive Australian trees. The aloes occur scattered across the veld amongst short coastal grasses.

Near Sherwood in the Port Elizabeth area

Aloe micracantha is found in good numbers in gently undulating coastal grassland near Sherwood, a suburb of Port Elizabeth. Much of the habitat has been invaded by Port Jackson willows (*Acacia saligna*). Fortunately biological control is having some effect on reducing the numbers of these plants. The aloe populations consist of widely scattered individuals. They are best represented in areas where the grass cover is relatively short and thin. They avoid the rocky outcrops since their very large, deep root systems are best accommodated in areas of deep, sandy soil.

The area was visited on 19 February 2004. Some plants had finished flowering but many were still in full flower. Aloes that had flowered had nearly all produced seed capsules, indicating that in this open grassland habitat the species is attractive to its pollinators. An indication of the current health of this coastal grassland was a very large colony of *Cyrtanthus obliquus* growing on stony, sandy outcrops in the vicinity of the *Aloe micracantha* plants. This bulbous species quickly disappears in degraded or over-grazed habitat, a situation prevalent on the recently extended Grahamstown common.

There are, however, a number of indications of impending habitat degradation. Kikuyu grass is becoming invasive along a track used for the dumping of garden refuse. The grass forms dense carpets along the side of the track and is starting to invade the surrounding veld. Much of the garden refuse decomposes but some building rubble is also dumped close to the aloes. South African towns have legislation in place to regulate dumping. In practice, however, there is too little manpower to ensure that dumping always takes place at designated sites.

Near Bridgemead

Bridgemead is a suburb of Port Elizabeth not very far, as the crow flies, from Sherwood. In times gone by the whole area would have consisted of uninterrupted gently undulating coastal grassland. *Aloe micracantha* is well represented near Bridgemead. The plants are still plentiful in short, open grassland. They tend to avoid areas with low *Leucadendron* scrub as well as low stony outcrops.

At present there is little, if any, invasive alien vegetation. This habitat consequently gives a good idea of what the distribution of these plants was like before the Eastern Cape became extensively colonised by alien plants. The only factor regulating the occurrence of *Aloe micracantha* are patches of *Leucadendron* scrub and areas where grass is naturally dense. These situations are usually the bottom of slight depressions, a microhabitat not favoured by *A. micracantha*.

The Bridgemead site was visited on 19 February 2004. Most aloes were found to be single-stemmed, but occasional plants were encountered with two or three flowering stems. One plant was very large, with 12 flowering stems. The aloes were mostly in full bloom, but some had flowered earlier and then had stems full of ripening capsules, suggesting a healthy presence of pollinators. The surroundings contained a number of indications of a diverse and rich grassland plant community. The orchid *Disa venusta* was well represented and several small bulbs were also present. These quickly disappear in degraded or over-grazed grassland.

Some problems facing this extensive population of *Aloe micracantha* in the future are urban expansion, dumping and invasion of the veld by alien vegetation.

Growth cycle and the role of fire

Aloe micracantha grows mainly in summer and flowers from January to March, with a peak in the second half of January and early February. The leaves wither in the early winter, typical of grass aloe species. The leaves die back

Right: An inflorescence of a plant of ***Aloe micracantha*** growing along the Mountain Drive near Grahamstown in the Eastern Cape. Plants in this area are set to benefit from the rehabilitation of the Featherstone Kloof, an area until recently smothered with alien trees.

mostly during dry winters, but winter may also be wet near Port Elizabeth and in the area to the south lying at the transition between winter and summer rainfall. A colony of plants near Joubertina had green leaves in late winter in a year of high winter rainfall, suggesting that the plants can remain evergreen when there is good winter rainfall.

Capsules of seeds ripen in March and April and are liberated into the environment at a time of lower temperatures and decreasing day length. Most seeds obtain their first chance to germinate in the rainy spring, about five to six months after they have been liberated from the ripe capsules. It is possible that *Aloe micracantha* seeds may germinate in habitat during mild, moist winters in the Port Elizabeth area. There are, however, no observations of this at present. Capsules are sometimes heavily parasitised by seed wasps, a situation encountered in many aloe species, including grass aloes, when they grow close together in large numbers.

Fire plays a significant role in keeping the veld clear from accumulations of dead grass. The most advantageous burns are those in healthy grassland that does not become too dense. These fires do not burn the veld with much ground heat and only the dead foliage of the aloes gets burnt. This type of habitat cleared by periodic fires is ideal for the germination of seeds and the development of seedlings.

Grassland invaded by aliens is quite a different matter when it comes to fires. Fires burn with a great amount of ground heat and smouldering branches and logs exacerbate the situation. Areas that are rehabilitated after clearing out and burning of alien plants would have to be repopulated with *Aloe micracantha* within a carefully managed veld rehabilitation programme. The Featherstone Kloof near Grahamstown could serve as a model for such a project since some plants of *A. micracantha* still grow there. These plants could be studied to ascertain the extent to which they are able to repopulate the rehabilitated environment with seedlings.

Status of wild populations

Aloe micracantha is rather sparse and rare around Grahamstown and much of the habitat suited to it is seriously threatened by invasive alien vegetation. The grasslands south and southwest of Port Elizabeth still have good representative populations of *A. micracantha*, but they are likely to dwindle over time as a result of the suburban expansion of Port Elizabeth, the spread of invasive plants and, to a lesser extent, the dumping of various forms of waste. It would be useful to identify areas where *A. micracantha* is still common and to encourage landowners to conserve the plants on their property. This could be carried out as part of a management plan for the preservation of Eastern Cape coastal grassland.

Cultivation, propagation and horticultural potential

Aloe micracantha has been successfully propagated by Welland Cowley, the owner of a specialist indigenous nursery in the Eastern Cape. Philip Crous, another Eastern Cape nurseryman living in Grahamstown, has also had good results propagating this aloe from seed. The account that follows is based on information supplied by these two gentlemen as well as experiences the author has had growing the plants in Johannesburg.

Aloe micracantha grows readily in containers filled with a well-drained soil mix, or in cool parts of a garden with a southern or eastern aspect. The plants grow in the summer but should also be given some water in winter as their natural habitat lies in the transition zone between winter and summer rainfall. In Johannesburg, the plants grow well in undisturbed, sloping, sandy, garden beds lightly shaded by shrubs and large trees.

Seeds should be collected in February and March when the capsules start to turn brown and begin to split open. Racemes with mature seed capsules can also be cut off the parent plant and allowed to ripen in a warm, dry place. These capsules soon open, liberating the seeds. Seeds germinate well when sown in a mixture consisting of four parts coarse river sand, three parts fine river sand, three parts sieved compost and two parts sieved loam. A small amount of bone meal should also be added to this mixture as a slow release fertiliser.

Seed trays should be filled with the sowing mix, which is best levelled off about 1 cm below the rim. The mixture should be well watered and allowed to settle. Treatment of the sowing mixture with a fungicide before sowing helps to prevent damping off. Seeds should be sown sparsely and broadly, with spaces of some 5–10 mm between them. They should be lightly covered with not more than 3 mm of fine river sand. The trays should be kept moist in an area with about 50 percent shade cover. Germination usually occurs between three and six weeks after sowing. Watering has to be carefully controlled at this stage, ensuring that the trays are not kept too damp. Over-watering usually results in attacks by damping-off fungi. The young plants can be left in the seed trays for up to six months and then transplanted into 12 cm pots in a well-drained soil mix.

This species can also readily be grown from cuttings struck in spring or summer. The cuttings should be removed from below ground level to ensure that good lengths of stem material are available for rooting. Cuttings should be left to dry in a well-ventilated place for about a week and then treated with a fungicide before they are planted. They should be struck in moist, coarse sand and positioned about 5 cm apart. If the cuttings are planted too close together there is insufficient air circulation and they are prone to rot. A rooting hormone stimulates root development, which usually starts within four to six weeks. The cuttings should be watered at intervals during the rooting process. They can be transplanted once they have developed their initial root system.

Philip Crous found that, in Grahamstown, seedlings were best left in their germination containers for at least nine months. By this time they had developed leaves 40–80 cm long. These aloes develop deep root systems and should be regularly transplanted into larger containers as they mature.

Plants left growing in 15 cm containers flowered for the first time in their fifth year. The aloes would have reached flowering size more rapidly had they been regularly potted on as their root systems developed.

Aloe micracantha makes a good container or garden subject and is very floriferous in cultivation. The plants are usually pest-free when grown in cultivation, but are sometimes attacked by woolly aphids.

16. *Aloe minima* var. *minima*

Aloe minima Baker var. *minima*

This is a plant of hilly and mountainous grassland. It is found mainly in Mpumalanga and KwaZulu-Natal where large tracts of its former grassland habitat have been destroyed by timber plantations. Most places where the plants currently occur are alongside afforestation areas, usually in rocky situations unsuitable for planting eucalypts and pines. In several areas that have not been afforested with aliens, the grassland habitat is heavily over-grazed and degraded. It is also burnt during the dry winter to stimulate spring and early summer grazing.

Grassland habitat around pine and eucalypt plantations is actively managed to prevent fires from destroying timber. *Aloe minima* var. *minima* has adapted to these artificial burning regimes in various ways, but the overall results appear to be a decline in the numbers of plants, over time, in the smaller populations. *A. minima* var. *minima* is one of the dwarf grass aloes that is likely to decline to the point of local extinctions in many parts of its present distribution range during the course of this century. It is least likely to survive in over-grazed, degraded grassland and several populations in this habitat have disappeared over the last two decades.

Reynolds (1969: 120) recognised *Aloe minima* var. *blyderivierensis*, said to differ from the typical *A. minima* var. *minima* in a few respects. *A. minima* var. *blyderivierensis* was described as having distichous leaves, slightly broader and more fleshy than those of *A. minima* var. *minima*, and also slightly larger flowers. None of these characters are constant and the typical form of *A. minima* frequently has distichous leaves. The only variable usually observed in modern populations of *A. minima* var. *blyderivierensis* is that the plants are much more gracile than those of typical *A. minima*. This may be as a result of significant inbreeding in the few small populations of *A. minima* var. *blyderivierensis* that remain. Most populations of this variety are isolated from one another by extensive afforestation and are certainly now genetically isolated from the typical *A. minima*. *A. minima* var. *blyderivierensis* is not recognised by Glen & Hardy (2000: 9), but no reasons are advanced in support of this decision. *A. minima* var. *blyderivierensis* is upheld in this work since there are still reasons for recognising its varietal status, even if small, and different from those originally proposed. These are discussed more fully in the chapter on *A. minima* var. *blyderivierensis*.

The view could also be taken that *Aloe minima* var. *blyderivierensis* is nothing more than a minor variant of *A. minima* that has become genetically isolated from the mainstream Mpumalanga populations, through afforestation. Plant taxonomists will have to deal with many such examples of fragmented populations when assessing the status of different species and varieties occurring across significantly altered natural habitats.

Aloe minima has been studied for nearly two decades near The Brook and Lochiel in east-central Mpumalanga. The study area has afforded the opportunity to record the lives of plants in areas heavily afforested with eucalypts and pines and also in over-grazed densely populated areas close by. These two habitats would have consisted of continuous mountainous grassland in former times, with none of the alterations that characterise the modern landscape.

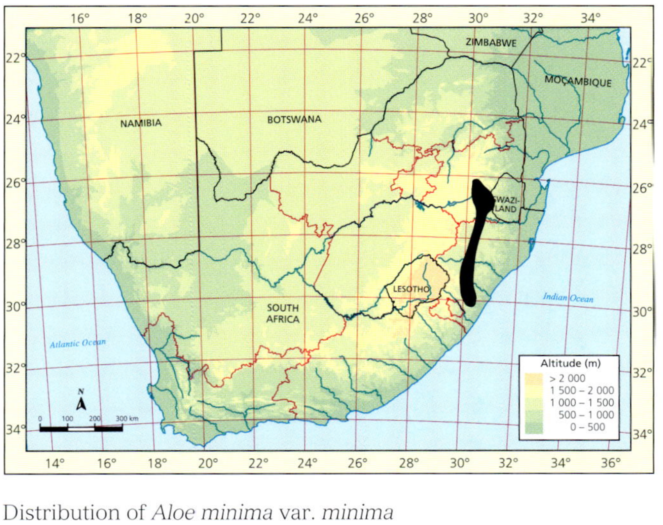

Distribution of *Aloe minima* var. *minima*

Near The Brook in east-central Mpumalanga

The countryside around The Brook once consisted of hilly grassland. From about the second half of the last century, the region has been progressively and extensively afforested with eucalypts, pines and wattles. Some parts of the countryside have small strips of grassland, only a few hectares in extent, hemmed in between plantations. In other areas there are virtually uninterrupted strips of grassland extending for many kilometres along the courses of seasonal streams. Extensive areas of rocky outcrops and sheets of exposed rock have also escaped afforestation. This territory often adjoins that along the streams.

Small areas of grassland between plantations tend to get burnt annually as part of fire prevention measures. The grassland along streams, in rocky hilly country, usually burns when the grass has become dense and thick at intervals of two to four years. The fires generally originate in the densely populated parts adjoining the alien tree plantations. They burn unevenly in the rocky convoluted grassland and some areas escape burning altogether if there are changes in the direction of the wind during a fire.

Plants are widely scattered in the grassland, usually occurring singly but sometimes in groups of about two to six plants where the grass is short and the ground rocky. The upper hill slopes and their crests are the most favoured habitat, with east- and southeast-facing slopes supporting most plants. The most robust plants, which are generally the most numerous, are found in deep soil in moist grassland, often in the vicinity of seepage areas. The smaller plants, which are sometimes only a little larger than *Aloe albida*, grow in rocky areas with sparse grass cover. The small size of these plants is related to the limited area they have available for developing their roots.

Soil depth is an important factor for determining the size of *Aloe minima* var. *minima* populations on the various hillsides where the plants are found. Deeper soil retains moisture for longest, particularly near seepage areas. These conditions are conducive to seed germination and development of seedlings. The rocky habitat dries out quickly and seedlings obtain the chance to develop only during growing seasons within a cycle of good rain.

Northeast of Lochiel in east-central Mpumalanga

Lochiel is not very far, as the crow flies, from The Brook. The region consists of hilly grassland densely settled with rural pastoralists and small-scale crop farmers. The habitat is heavily grazed and frequently burnt. *Aloe minima* var. *minima* grows together with *A. kniphofioides* in rocky grassland. Both plants have declined sharply in the past two decades and are now confined to the rockiest sites where they escape the heaviest grazing. The two aloes have largely disappeared from the erstwhile most suitable habitat consisting of moist grassland near drainage lines and around seepage areas. The remaining plants provide important sources of seeds for establishing more widespread communities of both grass aloes if rotational grazing is practised in future years.

Growth cycle and the role of fire

The hilly grassland in which *Aloe minima* var. *minima* grows receives most of its rain from November to March, mostly in the form of thundershowers. Mists are frequent in summer, providing the plants with the equivalent of light rain. Winters are cold, often accompanied by severe frost, and occasionally some rain or snow may fall. This, however, does not result in any leaf growth as *A. minima* var. *minima* is essentially a summer-growing aloe.

Leaves start to sprout and develop with the beginning of the main summer rains in late October and early November. They are usually fully developed just before the southern summer equinox. Flower buds develop in January and the peak of the flowering season is usually the first two weeks of February. Occasional plants may start to flower already in late January, usually as a result of heavy, consistent rainfall in October and November.

Seeds are liberated into the environment mostly during the last week of March and the first two weeks of April. five seeding plants were examined in a colony at The Brook on 10 April 2004, just as they started to liberate seeds. A total of 525 fertile seeds were counted in five seed heads. Flowering was overall poor in 2004 owing to dense grass cover but no seeds had been parasitised by seed chalcids. These diminutive wasps destroy many seeds some years when the aloe flowers well and has a good seed set. Late March and April is a time of the year when temperatures drop and dry autumn weather generally begins. Seeds lie dormant in the environment until the following summer when they have their first chance to germinate.

Fire is integral to the flowering cycle of *Aloe minima* var. *minima*. The short grassland becomes very dense during inter-fire years and flowering performance of the plants tapers off. Plants flower very well in late summer and early autumn after a grass fire the previous winter. The grass blades lightly shade the aloe leaves and developing flower buds the first year after a fire. In subsequent years, if the grass becomes very dense, only the most robust plants manage to elevate their flowering stems above the grass cover where they receive enough sunlight to flower.

The second season after a fire is usually the most favourable time for germination of seeds and development of seedlings of this grass aloe. The grass cover is becoming thicker but filters in the correct amount of light for seedling development. Seedlings germinating early in the summer are sufficiently robust at the end of the first growing season to resist destruction in a grass fire the following winter.

Conditions under which there is good flowering and effective pollination, followed by seed development without parasitism by seed chalcids, are rather rare. Added to this, is the fact that the veld must be burnt in at least the second season after seedlings have germinated to ensure that the reproductive cycle is completed under favourable conditions. These circumstances are rarely met, accounting for the fact that many populations of *A. minima* var. *minima* are small and widely scattered.

Status of wild populations

Aloe minima var. *minima* has a wide distribution in KwaZulu-Natal and Mpumalanga. Huge amounts of its former habitat have been afforested with aliens and other grassland areas where the plants grow, are becoming progressively more densely settled. The transition from cattle ranching to game farming east of Piet Retief in Mpumalanga and in parts of KwaZulu-Natal may benefit the plants and could save them from extinction in some areas. However, this grass aloe can be expected to continue its decline to the point of local extinctions during the course of this century owing to extensive habitat destruction and degradation.

Cultivation, propagation and horticultural potential

Aloe minima var. *minima* is difficult to maintain in pot culture for long periods of time. The plants require a well-drained growing medium consisting of a third fine silt, a third sieved sandy soil and a third finely sieved, very well-rotted leaf mould. Plants should be well watered during the period of active growth, which is November to early March in the southern hemisphere.

This aloe grows well in positions receiving only morning sunlight or light dappled shade. These conditions approximate those in the natural grassland habitat. The plants should be kept bone-dry in shaded positions in winter. They should be watered lightly and occasionally in late winter and early spring, a few months in advance of the main growing season. The plants grow best in earthenware containers absorbing excess moisture. They thrive in containers known locally in South Africa as moss pots. These containers, made by a specialist potter in Gauteng, consist of very porous, gritty clay that accumulates large concentrations of moss as the years pass. Care has to be taken when plants are grown in containers to ensure that the base of the pots are kept free of ants. These insects often introduce root mealybug to *Aloe minima* var. *minima*, generally killing the plants if left untreated.

Propagation is from seed. Seeds should be lightly and evenly scattered across the surface of the soil. Not all seeds of *Aloe minima* var. *minima* germinate the first year. Some seeds germinate at the beginning of the second summer when watering is resumed once more at the end of the winter dormancy. The mixture described for the cultivation of this species also provides an excellent germination medium. Large, deep seed trays are ideal for this aloe as they

Right: ***Aloe minima* var. *minima*** in its short grassland habitat near The Brook in Mpumalanga. The various phases of bud, flower and fruit development take place in February, March and early April.

allow the seedlings to develop their thickened roots unimpeded by shallow soil. Good drainage at the bottom of the seedling tray is essential. The germinating seeds and seedlings have to be kept moist. Care has to be taken to ensure that seedlings do not damp off. Seeds sown early in spring mostly germinate before the heat of summer when damping-off fungi are prevalent. It is advisable to spray the seed trays with a fungicide from the time the seeds are sown until the seedlings are a few weeks old. Seedlings are best transplanted at the end of the second growing season.

Aloe minima var. *minima* is an interesting horticultural subject, but rather difficult to grow well in cultivation. If plants are required for *ex situ* conservation programmes, they are probably best suited to cultivation in a grassland veld garden in regions of high-altitude grasslands.

17. *Aloe minima* var. *blyderivierensis*

Aloe minima Baker var. *blyderivierensis* (Groenew.) Reynolds

Aloe minima var. *blyderivierensis* is currently found in a few places in the Graskop and Pilgrim's Rest areas of Mpumalanga. Most of this region is densely afforested with eucalypt and pine plantations and as a result the grassland habitat is very limited and fragmented. The plants still occur at one of the localities recorded by Reynolds (1969: 120). This large stretch of grassland has escaped afforestation and is in relatively pristine condition. Much of the remaining veld in the area has been degraded to various degrees by invasive alien weeds and trees.

Most authors who have written about *Aloe minima* var. *blyderivierensis* comment on its close similarity to *A. minima* var. *minima* (Jeppe 1969: 127; Bornman & Hardy 1971: 9; Van Wyk & Smith 1996: 278). The variety *blyderivierensis* was reduced to synonymy under *A. minima* by Glen & Hardy (2000: 9) in their revision of *Aloe*. No reasons were, however, given for this decision.

Aloe minima var. *blyderivierensis* is upheld in this work for several reasons. The habitat where the plants occur is so fragmented by plantations that it is impossible to study regional variation in these aloes. Ideally, representative populations between Graskop and Lydenburg, where the typical *A. minima* is found, should be examined to reach conclusions about the status of *A. minima* var. *blyderivierensis*. There is a strong international development at present aimed at recognising and protecting biodiversity. This movement may precipitate more widespread use of subspecies and variety as concepts for detailed organisation of knowledge in plant classifications.

Aloe minima var. *blyderivierensis* was said to differ from *A. minima* var. *minima* in having fewer, slightly broader, distichous, more fleshy leaves and somewhat larger flowers (Reynolds 1969: 120). The data collected for this book indicated that the leaves can be either rosulate or distichous where groups of these aloes are growing close together. The leaves are mostly distichous, however, when the plants grow singly, widely scattered in hilly grassland. There are also a few other minor differences. *A. minima* var. *blyderivierensis* grows in short grassland as well as on rocky outcrops with little grass cover. The aloes often form small clumps in these situations. *A. minima* var. *minima* is usually not a rock outcrop plant, even where it occurs in areas where this type of habitat is extensive. *A. minima* var. *blyderivierensis* generally has a slightly earlier flowering season than *A. minima* var. *minima*. One population near Graskop studied for about a decade consistently comes into flower in the second half of January. The plants growing in this scattered colony are a good example of the current habitat that remains in the Graskop area.

Right: ***Aloe minima* var. *blyderivierensis*** in grassland near Graskop in Mpumalanga. The few populations that still exist in the area are all hemmed in between eucalypt and pine plantations.

Near Graskop

Various segments of grassland near Graskop have not been forested with eucalypts and pines, either in the vicinity of the waterfalls or viewpoints over the escarpment, for which the area is renowned, or where the ground is too rocky for the establishment of plantations. The locality where the aloes were studied consists of hilly grassland with steep slopes on the northern, northwestern and western aspects. The eastern and southern slopes are afforested with eucalypts and pines.

The aloes grow in three distinct types of habitat—in short but dense grass, on stony ground with more limited grass cover, and on rocky outcrops that include sheets of exposed rock harbouring shallow patches of soil. The plants are widely scattered in grassland and usually occur singly. They grow closer to one another on stony ground and sometimes occur in small groups of two to four aloes. The rock outcrops have a limited amount of short grass cover but are home to extensive clumps of moss. *Aloe minima* var. *blyderivierensis* is most common in these situations and often forms clumps of three to 15 individual plants. Many seedlings and young plants of different ages are found on the rocky outcrops.

The soil in the area has a high clay content and is consequently moisture-retentive. The sheets of rock often lie within seepage areas, ensuring that the aloes obtain a regular supply of water in summer. The area receives regular mists in summer, particularly from December to February which is the peak of the growing and flowering season.

The veld in which the aloes occur is joined to other grassland areas by corridors of intact indigenous vegetation. There were no livestock grazing the veld at the time the data for this book were collected and as a result the aloes were well distributed in open grassland. The veld is, however, in the initial stages of invasion by black wattles (*Acacia mearnsii*) and pines (*Pinus patula*). In recent times, the trees have been chopped out to arrest the process but it will have to be sustained if the integrity and richness of this strip of grassland is to be maintained in the future.

This population of *Aloe minima* var. *blyderivierensis* is likely to comprise only a fragment of what were once much larger groups of plants in the general area, prior to afforestation with pines. The aloes appear to be healthy, but the gene pool is limited to a very small area of grassland where the plants are currently found. The aloes flower and seed well. This colony of plants is an exception to the general rule of pollinators invariably being reduced in numbers and variety in regions extensively afforested with timber. No plants of *A. minima* var. *blyderivierensis* were, however, found in small stretches of grassland in the Graskop area and as a result there was no possibility of comparing pollination rates between plants growing in the small and large remaining patches of grassland.

Growth cycle and the role of fire

The plants of *Aloe minima* var. *blyderivierensis* near Graskop grow in a fire prevention area. Despite this, fires occur fairly regularly in areas with large patches of grassland. The occurrence of fires is linked to large numbers of itinerant and resident people in the Graskop area throughout the year. Fire plays an important role in keeping the areas of short grassland free from moribund vegetation. The aloes always flower at their best after the veld has been burnt the previous winter. The plants growing on rocky

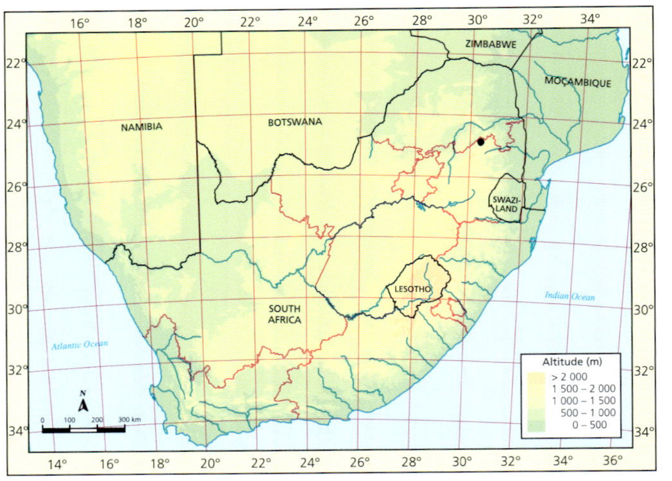

Distribution of *Aloe minima* var. *blyderivierensis*

outcrops flower well each year where the grass cover is sparse. The occasional lack of grass fires for a few successive years has the effect of restricting the opportunities for widespread cross-pollination in the habitat at large.

The aloes growing in grassland restrict their optimum flowering and seed production to seasons following a winter grass fire. At these times the grass cover is sufficient to shade newly germinated seedlings and also to permit free air circulation necessary for the development of the young plants.

The aloes start to sprout in November once the first heavy mists and rains of the summer begin. Warm spring weather in September and October, which are usually dry months, does not usually cause the aloes to initiate leaf growth if it does not rain. The leaves continue to grow throughout November and December. Flower bud formation is initiated in early December and is usually well under way at the time of the southern summer equinox. The plants flower mainly in the second half of January, extending into early February when the rains begin late in the summer.

Seeds are dispersed from early to mid-March, mostly by wind and water runoff after rain. Conditions are often damp during March and early April and some seeds obtain their first chance to germinate at this time. The rest of the seed germinates at the beginning of the following summer.

Status of wild populations

The status of *Aloe minima* var. *blyderivierensis* is precarious owing to the extensive destruction of its habitat by alien timber plantations. The plants are unlikely to survive in the long term unless the areas where they grow are actively managed. Management programmes will have to ensure that the veld is kept free of invasive alien plants and that it continues to be burnt every few years. Timber companies could play a key role in the conservation of these plants.

If the varietal status of these aloes is no longer upheld, their disappearance from the Graskop grasslands is of little overall significance.

Cultivation, propagation and horticultural potential

Aloe minima var. *blyderivierensis* is a difficult species to maintain in cultivation over the long term. The plants should be grown in a well-drained medium and preferably cultivated in earthenware containers. A good cultivation mixture consists of one third coarse gravelly sand, one third sieved sandy soil and one third very well-rotted compost or leaf mould. The plants require plenty of free air circulation and grow best in positions of morning sunlight. They should be well watered from early summer until early autumn. The aloes should be kept dry during winter dormancy. Occasional watering towards the end of winter and in the early spring improves the condition of the plants for the beginning of their growth cycle in the early summer.

Seed yields are improved when the flowers are hand-pollinated, otherwise little seed is set in cultivation. Seeds should be sown in spring or early summer. They should be evenly scattered across the surface of the soil and covered by a shallow layer of the mixture used for germination, not more than 1–2 cm deep. Seeds can be sown in any well-drained, gritty mixture in deep seed trays or pots. The containers should be placed in lightly dappled shade during germination and for the duration of the first growing season. In the second year they can be moved to positions of morning sunlight. Seedlings have usually developed sufficiently for transplanting by the end of their second growing season. They are, however, best transplanted in late spring just before their summer growing season.

18. *Aloe modesta*

Aloe modesta Reynolds

Aloe modesta was described in 1956 by Gilbert Reynolds, doyen of South African aloes. At the time the plants were known from two localities in Mpumalanga, one near Dullstroom, the other near Wakkerstroom. Both these habitats are in high-altitude, short grassland experiencing frequent rain and dense mists in summer. The species is easily overlooked unless in flower or seed. It is rarely recorded for two reasons—its short flowering and seeding period restricted to February and March and the fact that it is rare. Many grassland areas in KwaZulu-Natal and Mpumalanga were searched at the appropriate time of the year, but *A. modesta* was not found in any of them.

Aloe modesta is the only species of *Aloe* with scented flowers on the African continent. The scent resembles that of cinnamon mixed with lemon juice. The flowers are strongly fragrant, particularly during the late part of the morning and in the early afternoon.

It is one of three species in South Africa with a bulb-like swelling, the other two being *Aloe inconspicua* and *A. kniphofioides*. The latter grows sympatrically with *A. modesta* on a hilly escarpment near Dirkiesdorp in Mpumalanga. It flowers in late October and early November, at the beginning of the summer rains.

Aloe modesta is very susceptible to degradation of its habitat, which occurs through over-burning, over-grazing and degradation of the grassland by weeds and introduced invasive Australian trees. Black wattles (*Acacia mearnsii*) rapidly transform the grassland habitat where this grass aloe grows. They initially invade the stream banks and seepage areas, resulting in lower moisture levels in the grassland frequented by the aloes. Once the trees become numerous along drainage lines, they begin to invade the grassland directly as their seeds start to accumulate in the environment.

The aloes have been studied for some 15 years where they grow in short grassland on an escarpment near Dirkiesdorp, a small town in southeastern Mpumalanga.

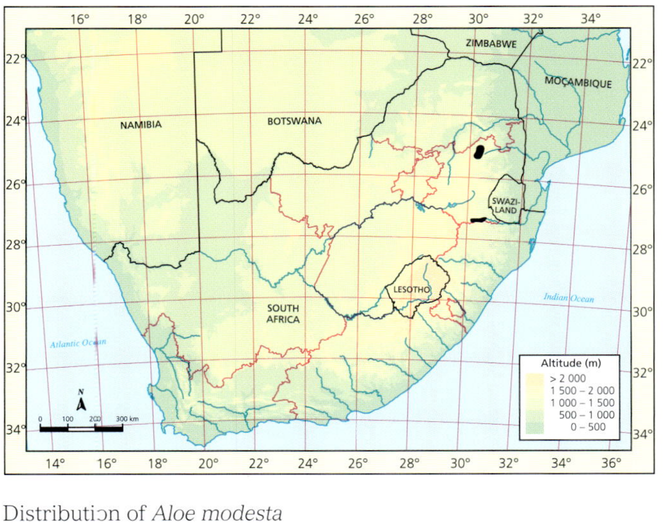

Distribution of *Aloe modesta*

Colonies near Dirkiesdorp in Mpumalanga

Aloe modesta occurs near Dirkiesdorp on the summit of a range of hills at an altitude of about 1 800 m. The aloes grow in short grassland and are best represented in rocky areas where the grass cover is shortest. The summit of the escarpment consists of gently undulating hills, the ground sloping off very steeply to the south and southeast. The aloes are found mostly within an area of 400 m from the edge of the escarpment and are rarely encountered in the adjacent flatter grassland.

The plants occupy three types of habitat where the most moisture is retained after rain—short grassland amongst rocks on the lip of the escarpment, in thin grass cover around the edges of sheets of exposed dolerite, and in areas of short, dense grass often near the perimeters of seepage areas. Some of the habitat has been degraded through erosion brought about by heavy grazing in the past. A more serious problem is the transformation of the grassland in the general area by invasive black wattles. The aloes do not grow in the seepage areas, which are marshy throughout the summer. They are also absent from areas with the tallest grass.

The locality where the plants occur has a high summer rainfall and also lies in a mist belt. The mist condenses to form light, penetrating rain and this form of precipitation is at its heaviest around the lip of the escarpment where the aloes are most plentiful.

In some years the veld is burnt during dry weather in summer. This has the effect of killing off the leaves of the *Aloe modesta* plants and also any buds or flowers present at the time of a fire. Summer fires are deliberately started to stimulate autumn grazing. The grass is not always flammable in summer, particularly if the veld was burnt the previous winter, as there is very little dry grass to fuel the fire. When this practice was first noticed, during the course of the research on *A. modesta*, it was thought that there would be a negative impact on the regeneration of the plants. The effects of summer burning are, however, less than expected. The main reason is that flowering is best after a winter grass fire. This is the time at which summer grass fires are least likely owing to the abundance of fresh, relatively inflammable grass cover.

Growth cycle and the role of fire

Aloe modesta starts its growth cycle in November with the beginning of the main summer rains and mists, sprouting leaves soon after the first rain. Leaf growth occurs mostly in November and December. Flower buds start to form in early January and the plants start to flower in late January and early February. The peak of the flowering season is usually the first two weeks of February.

The aloes produce their seed capsules in the second half of March and seeds are liberated into the environment in late March and early April. Seeds are dispersed by wind and water runoff after rain. They are liberated from the parent plants at a time when temperatures start to drop with the approach of autumn. The mist and rain may persist until well into April, but most seeds do not germinate until the following October or November when the summer rains begin. *Aloe modesta* is dormant for the duration of the dry winter. The leaves shrivel completely in late April or early May, a time when the early frosts of the winter start to occur.

Aloe modesta is fully dependent on fire for its growth cycle and reproduction. Fire removes accumulations of dead grass from the veld and opens up the habitat, creating suitable conditions for seeds to germinate. Mass flowering occurs after a grass fire the previous winter and the flowering performance of the plants tapers off if the veld remains unburnt for several years in succession. During these times, flowering is restricted to the aloes that grow in places with the least cover, such as areas of thin grass around the perimeters of exposed rock.

Status of wild populations

This is a rare species absent from habitat that appears to be ideally suited to its requirements. A large part of potential habitat in Mpumalanga has been afforested with timber plantations, but there is no information as to whether or not *Aloe modesta* used to occur in these areas. Large sections of the mountainous grassland near Dirkiesdorp have been transformed by black wattles. The wattles also occur in a substantial proportion of the grassland immediately east of the Dirkiesdorp *A. modesta* populations and some of the veld have already been transformed into dense stands of wattles. The process has been accelerated by areas of disturbed soil that are rapidly colonised by these invasive trees. As the seed bank increases, the wattles start to invade the grassland in larger numbers.

Aloe modesta is best preserved by drawing the attention of landowners to its presence on their property. They should be encouraged to keep the veld clear of invasive weeds and trees and to ensure that the grassland is burnt every few years. The plants of *A. modesta* growing near Dirkiesdorp seed well and the pollination of the plants has not been affected by stock farming, which is the main form of land use in the region. There are programmes in place for the control of black wattles and other invasive introduced plants, but they will have to be sustained for many decades if the long-term survival of rare grassland plants such as *A. modesta* is to be secured.

Cultivation, propagation and horticultural potential

Aloe modesta requires a well-drained, peaty soil to grow well. It is best grown as a container plant. A good cultivation mixture consists of one third silt or gravelly sand, one third sieved loam or sandy soil, and one third finely sieved peat or very well-rotted compost.

The plants should be grown in clay pots absorbing excess moisture after watering. These aloes are best cultivated in morning sunlight or very lightly dappled shade. They should be well watered throughout the summer. Watering should be gradually reduced after the plants have finished flowering and then continued rather sparingly until the leaves start to show signs of dying back at the beginning of autumn. Plants of *Aloe modesta* should be occasionally and lightly watered during winter but otherwise kept completely dry.

Right: ***Aloe modesta*** grows amongst short grass and is usually at its most frequent on stony outcrops or sheets of exposed rock where it escapes competition from taller grasses and robust, deep-seated bulbs. These plants were flowering on an escarpment near Dirkiesdorp in Mpumalanga in early February.

The aloes are best propagated from seeds, which should be scattered over the surface of the germination mixture and covered with a thin layer of soil. The plants germinate well in the same soil mix recommended for the cultivation of this species. Large seed trays or earthenware pots are best for growing the aloes from seed. They should be placed in positions that have day-long dappled shade or a few hours of direct sunlight in the morning. Seeds should be sown in early summer. They germinate best when exposed to rain, like the seeds of all the South African grass aloe species.

Seedlings are very prone to attack by damping-off fungi, particularly if the weather is humid and rainy. The young plants should be regularly sprayed with a fungicide for about two weeks if any signs of damping off are noticed. The aloes are ready for transplanting as soon as they have developed small bulbs, usually at the end of their second growing season. *Aloe modesta* is regularly attacked by root mealybug. Withering of some of the leaves as well as poor leaf development is usually indicative of infestations with this pest.

The plants produce seed sparingly in cultivation, but the yield can be improved by hand pollination. Seeds should be kept in paper envelopes once they have been set, in late autumn. They should be sown at the beginning of the following summer.

Aloe modesta sometimes flowers much earlier in cultivation than in its natural habitat and may carry well-formed flower buds by late December. It is really a subject for the specialist aloe cultivator. Apart from the attractive and unusual character of strongly scented flowers, it is rather inconspicuous and drab when compared with other showy grass aloe species.

19. *Aloe myriacantha*

Aloe myriacantha (Haw.) Schult. & Schult.f.

Aloe myriacantha has been recorded mainly from the Eastern Cape, particularly the Albany District, around Grahamstown. There are also records of the species from KwaZulu-Natal, with some recent sightings of what appears to be this species from the Port Edward and Oribi Gorge areas in the southwestern part of the province. Like all dwarf grass aloes, the plants are very inconspicuous when they are not in flower. They occur singly or in small scattered groups of two to six or a few more. The small numbers of aloes in a given area, and also their absence from large parts of suitable habitat, imply that the species may well be overlooked in some regions.

The habitat in the Albany District consists of hilly grassland, severely invaded by alien vegetation and heavily grazed in most areas. Habitat alteration and degradation has existed in the area to varying degrees for over a century. It is consequently an ideal area in which to conduct research on this grass aloe, which is particularly susceptible to changes in its habitat.

Populations were visited and studied at the following localities: the Mountain Drive immediately southwest of Grahamstown, the Featherstone Kloof, the newly extended common to the east of the town, the railway-line reserve between Coldspring and Highlands and the adjacent farmland southwest of Grahamstown, and the gently undulating hills to the northwest of the town. These areas are characterised by different combinations of pressures on the habitats, affording ideal opportunities to assess the status of the aloes in their modern environment. The field research was greatly enhanced by the contributions of Charl Malan of Grahamstown and David Cumming of Bathurst. These two gentlemen have a thorough field knowledge of the Albany District. David Cumming has located several groups of *Aloe myriacantha* in different places around Grahamstown and north of Bathurst.

Plants in the grassy elevated areas of the Albany District have evolved with grazing by encamped domestic stock for well over a century. More recently, with the eastward extension of the Grahamstown common, managed farmland grazing has given way to communal grazing, with a wider range of domestic stock. Farmland is grazed mainly by cattle and sheep whereas the common is grazed by cattle, sheep, goats, horses and donkeys. The vegetation on the expropriated farms, which now comprise the extended common, are undergoing changes not encountered when the area consisted of farmland.

The Featherstone Kloof and surrounding area was largely covered with invasive aliens in the 1960s and 1970s. Huge parts of the habitat, as well as large tracts of land to the west between Coldspring and Highlands, were extensively invaded by black wattles (*Acacia mearnsii*), Port Jackson

Right: ***Aloe myriacantha*** in flower on the hills facing Grahamstown east of the Featherstone Kloof in the Eastern Cape. The site had been extensively cleared of dense tracts of invasive alien vegetation.

J Condy

willows (*A. saligna*) and long-leaved wattles (*A. longifolia*) as well as eucalypts and pines. This has resulted in the destruction of much suitable habitat probably occupied by scattered groups of *Aloe myriacantha* in former times. The alien vegetation has now been chopped down and burnt as part of the Featherstone Kloof Rehabilitation Project. This area is now in the process of reversion to its former grassland status and provides an excellent opportunity to assess the adaptation of the two grass aloe species occurring here, *A. myriacantha* and *A. micracantha*, to the newly created habitats.

The intricate processes at work in the Albany grasslands are discussed in more detail below, with reference to specific habitats.

Between Coldspring and Highlands, west of Grahamstown

The area between Coldspring and Highlands consists of hilly, rocky grassland with deeply incised valleys. This region is really an eastern extension of the Suurberge and lies at an altitude of mostly 700–800 m. The Grahamstown–Alicedale railway line runs along the summit of the hills. The associated railway-line reserve plays a significant role in the preservation of *Aloe myriacantha* in the modern environment. The railway line was completed well over a century ago, in 1879. The plant communities in the reserve are old and long-established. They have remained largely ungrazed but the grassland is burnt from time to time, ridding the veld of moribund grasses and herbs.

The surrounding farmland is well stocked with sheep and cattle. Sheep farming has been practised in the Suurberg area since about the late 1830s. Veld that has been heavily stocked for the last 150 years and more resembles the hedged-in, closely grazed pastures of the English countryside. Grassland plants that can withstand close cropping by domestic stock are abundant in these areas, particularly acaulescent caudiciforms. Other plants that cannot withstand heavy grazing pressures, such as *Aloe myriacantha*, have long disappeared.

In the area between Coldspring and Highlands *Aloe myriacantha* is largely a railway-line reserve plant. It is also very sparsely found amongst short grass on rocky outcrops on some farms in the highest areas with the rockiest grassland habitat. The railway-line reserve is variable in width, preserving only a tiny fraction of the grassland habitat. *A. myriacantha* is found in grassy, rocky areas of the reserve.

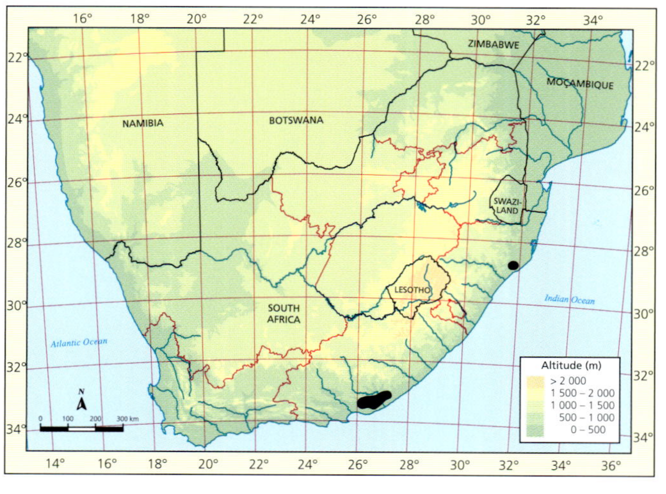

Distribution of *Aloe myriacantha*

The ideal habitat consists of scattered stones with tufts of short grass. The plants do not grow on rocky outcrops that are devoid of grass, and also not in patches of dense grass and open scrub. The suitable habitat available to the aloes is very limited, accounting for the fact that groups of plants are usually few and close together.

A search of the farmland on either side of the railway-line reserve occupied by *Aloe myriacantha*, usually yields the disappointing result of an absence of plants. These hillsides would certainly harbour some *A. myriacantha* were it not for grazing pressures. The aloes have a deep root system, making it unsuited to growth on rocky outcrops with sparse grass cover. The success of some grass aloe species is undoubtedly their ability to grow amongst short grass on rocky outcrops. This is typified by *A. saundersiae*, today common only on outcrops in the heavily grazed Nkandla area of KwaZulu-Natal.

The railway-line reserve habitat has the disadvantage that fire is the only natural process that eliminates dead grasses and herbs. The grass may become very thick between fires and when the veld does get burnt, the results are a deep, hot burn owing to the accumulation of dead grasses. Areas that are lightly grazed tend to have thinner grass cover and burn more lightly in veld fires. The populations along the railway-line reserve tend to consist mostly of mature plants since the dense cover in most areas does not afford seeds the chance to germinate and develop.

In one area immediately east of Coldspring, an *Aloe myriacantha* population has been destroyed by grazing within the last decade. Broad parts of the railway-line reserve are now grazed, particularly where pasture is sparse owing to extensive forests of alien Australian vegetation. The railway-line reserve is no longer well maintained and the fences have rusted away in several places between Coldspring and Highlands. This may open up large parts of the reserve to grazing by domestic stock, which would see the demise of *A. myriacantha* in this important microhabitat.

The Grahamstown Common

The expanding population of Grahamstown has resulted in a recent extension to the common. This has occurred via expropriation of farms to the east of the town. Some of these farms did not consist of ideal grazing lands and were consequently rather under-stocked. *Aloe myriacantha* used to occur in this area in typical small, scattered groups of plants. Though never common, the species was noted by several botanically minded people shortly after the common was expanded.

David Cumming knew of one population on the extended common close to the secondary road to Manly flats. A thorough search by David and the author in February 2004, when the species should have been about to flower, indicated that this population had disappeared. The area is currently heavily grazed and in the process of losing indigenous species susceptible to repeated trampling such as *Aloe myriacantha*.

Habitat immediately north of Grahamstown

The area to the immediate north of Grahamstown consists of some well preserved grassland close to the industrial sector of the town. The rocky, gently sloping hills contain patches of open grassland with isolated patches of typical

Eastern Cape bush. *Aloe myriacantha* occurs in small numbers of individual plants growing mostly a few metres away from one another. No small plants were observed, which typifies populations of this species around Grahamstown. There is likely to be a high wastage of the limited amount of seed that is set and, moreover, only small parts of the area are suited to the requirements of the aloe. Most of the surroundings contain grass and scrub too tall to house the aloes and the soil depth on the numerous pebbly outcrops is too shallow for the plants to establish themselves. The area is not grazed, except perhaps by an occasional antelope, and periodic fires thin out the grassland in winter.

The Mountain Drive above the Featherstone Kloof

The Featherstone Kloof Rehabilitation Project is a joint venture between the Albany Working for Water Project, the Rhodes University Institute for Water Research and the Grahamstown Municipality. This environmental rehabilitation project has been funded by the Department of Water Affairs and Forestry and the Donaldson Trust. The long-term objectives of the project are water security. Local human resources have been used for planning and effecting the removal of invasive alien vegetation.

The grassland that existed at the time Grahamstown was founded can be restored, even if in a slightly altered form. *Aloe myriacantha* was probably quite widespread along the Mountain Drive before the establishment of invasive aliens. The plants that exist now are those that were able to survive for many decades under pines, eucalypts, Port Jackson willows and long-leaved wattles. In many cases this has been for at least four decades and probably much longer.

These plants have grown normally and flowered in the early autumn once the alien vegetation has been felled and removed. The aloe illustrated was a specimen that used to grow in a eucalypt plantation before the rehabilitation of the Featherstone Kloof. It will be interesting to see how both *Aloe myriacantha* and *A. micracantha* take their final place amongst the flora in the resurrected grasslands of the Mountain Drive and the Featherstone Kloof in future decades.

The other grass aloe species that survives in alien tree plantations for many decades is *Aloe kniphofioides*. Once plantations of pines are felled near Piet Retief in Mpumalanga and the debris and the pine needles on the forest floor burnt, large numbers of *A. kniphofioides* come into flower in early summer.

The newly established grasslands of the Featherstone Kloof and surroundings are too young for an assessment of the status of many of the more sensitive plants such as *Aloe myriacantha*. There is also some indication that alien vegetation is once again starting to establish itself via a seed bank that is likely to remain viable in the environment for a long time. The Featherstone Kloof Project may well serve as a model for the preservation of some of the grassland of the Albany District, which is becoming progressively encroached by invasive aliens.

Growth cycle and the role of fire

Grahamstown experiences most rainfall in spring, September and October, and to a lesser extent in the autumn, February and March. Rain may, however, fall at any time of the year and may also be quite heavy in winter some years. *Aloe myriacantha* grows mostly in the rainy summer months but is usually dormant in habitat in winter. Even if rain does fall, the low temperatures are not conducive to the growth of foliage. The species is dormant in late autumn and winter.

The autumn rains are often accompanied by cloudy, misty weather, even in the absence of conventional rainfall. The mist is sometimes sufficiently dense to fall as soft, penetrating rain. This source of moisture certainly augments the moisture the plants receive at flowering time. A spurt of leaf growth occurs with the autumn rains and flower buds and stems are well developed in the second half of February. The flowering season is consistently late February and early March. A photograph in Reynolds (1969: 116) shows a plant in flower on 3 March 1935 at Coldspring. Plants observed close by in the railway-line reserve on 17 February 2004 were about two weeks away from peak flowering, an indication that the flowering season of this species has been remarkably consistent over the years.

Seeds are liberated into the environment in April at a time when it is too cool for them to germinate. They lie dormant until the rainy season the following spring when they have their first opportunity to sprout.

Fire plays a significant role in keeping the habitat clear of accumulations of moribund vegetation. It is most effective in veld without large accumulations of dead grass and which is consequently only lightly burnt. Deep, hot burns in which large amounts of flammable material get ignited are detrimental to the survival of the plants.

Seeds are produced every year, since the species flowers regularly from one year to the next. They are likely to germinate successfully only for the first few seasons after fire until the grass becomes too dense. These conditions partly explain the small size and scattered nature of the populations.

Status of wild populations

Aloe myriacantha may well be more widespread in the remnants of the coastal grasslands of KwaZulu-Natal than present records indicate and the same situation may exist in the Transkei. The Albany habitat has been altered to such an extent over the last two centuries that the plants could decline to the point of extinction in this area. The various causes of this decline identified in this chapter, could lay the basis for future studies on the species in the Albany District.

Cultivation, propagation and horticultural potential

Two Eastern Cape plant enthusiasts, Philip Crous of Grahamstown and David Cumming of Bathurst, have good experience with cultivating and propagating *Aloe myriacantha*. The discussion that follows relates their experiences growing this species and also the way in which these aloes respond to cultivation in Johannesburg. The plants adapt differently to cultivation in the high-altitude interior of South Africa, in comparison to their native eastern Cape coastal grasslands.

Both Eastern Cape growers recommend that *Aloe myriacantha* should be cultivated as a container subject on account of its small size. It is furthermore difficult to find a situation in the garden that is suited to growing the aloes and they rot if they receive too much moisture. David Cumming grows the plants in a well-drained mixture consisting of coarse sand, perlite and compost. They are housed under shade cloth and are watered by rain. Artificial watering is used only under very dry conditions.

In Johannesburg, *Aloe myriacantha* is grown in an open sandy bed amongst indigenous bulbs and *A. vossii*. The plants receive about half a day's sunlight. The aloes grow under these conditions and also under shade cloth, but they rarely flower under either of these conditions.

Propagation is mainly from seed. David Cumming and Philip Crous either plant seeds shortly after harvesting or store them in a refrigerator and then sow them from the end of August, at the end of winter. David uses a sowing mixture consisting of two parts Canadian or German peat moss and one part vermiculite. Philip uses a seedling mix consisting of equal amounts of coarse river sand, sieved compost and fine loamy soil. He uses 12.5 cm plastic pots for germination purposes. A good layer of stones is packed at the bottom of the pots to ensure adequate drainage. Seeds are evenly scattered across the surface of the germination mix and just covered. Philip sows 10 to 12 seeds in each 12.5 cm pot. He finds that germination takes three to eight weeks. David finds that with freshly harvested seeds, germination begins a few days after sowing.

The containers used for sowing should be watered regularly but sparingly throughout the germination period and afterwards while the young plants are developing. Seedlings should be transplanted when they have started to mature and produced about four leaves. The young plants benefit from an application of liquid fertiliser.

In Bathurst, plants of *Aloe myriacantha* rarely die back during the winter dormancy. They die back completely, however, under cultivated conditions in Johannesburg in winter. The plants should be kept moist throughout the summer growing period. They may be watered lightly but occasionally in coastal areas in winter. The aloes should be kept dry in winter when grown in the frosty South African interior.

The aloes can be induced to produce side shoots if their stem material is pruned. The suckers can be removed from the parent plants and struck in moist coarse gravel. This method of propagation is not recommended as the species is readily grown from seed.

Aloe myriacantha is not a showy species and also difficult to cultivate. It is therefore unlikely to appeal to anyone other than the committed cultivator of rare aloes.

20. *Aloe nubigena*

Aloe nubigena Groenew.

Aloe nubigena is a distinctive species found along the Drakensberg Escarpment mist belt around Graskop in Mpumalanga. The plants usually grow near the summits of tall precipices and occupy two distinct habitats. They occur in scattered groups on ledges of sheer rock faces where they grow in clumps of moss, or amongst short grass and moss-strewn boulders at the summit of the precipices. Most of the aloes occur on the upper third of the rock faces where the greatest amount of habitat is available to them and also where frequent summer mists condense as fine drizzle.

The habitat occupied by *Aloe nubigena* is similar to that of *A. thompsoniae* on the Wolkberg and *A. soutpansbergensis* on the Soutpansberg. All three species have adapted to habitats that are saturated for several days in succession in summer.

Much of the grassland on the mountain summits near Graskop is forested with eucalypts and pines. These plants as well as black wattles (*Acacia mearnsii*) have degraded much of the cliff-top habitat along the escarpment. The Graskop area has some of the most spectacular mountain escarpment scenery in South Africa, with many waterfalls and indigenous forests. Much of the general area where the aloes are found has been opened up to tourism and hiking. This accounts for a large number of people walking over the cliff-top habitat, disturbing the fragile mossy habitat where *Aloe nubigena* is found. The cliff-top colonies have declined over the last few decades, particularly around well-known waterfalls and viewpoint areas that attract large numbers of visitors. Most plants of *A. nubigena* in these places occur on cliff ledges and the species is still found in good numbers on mossy cliff-top boulders away from tourist attractions.

Aloe nubigena depends on high rainfall and two other sources of moisture for its existence—dense mists, which occur regularly in summer, and irrigation from cliff-top seepage areas. These seepages have been degraded by the vast plantations that have replaced the high-altitude grassland and consequently the aloes are rather scarce in mountaintop habitat where they were once more frequent. *A. nubigena* is found largely on rock faces with an eastern or southern aspect. The species was not located on any of the sunny north- and west-facing cliffs while the data for this book were being collected. North- and west-facing precipices are frequent around Graskop and some famous waterfalls, such as the Lisbon Falls, cascade over these cliffs.

This species has one of the longest flowering periods of all the South African grass aloes, from November to late March or early April. This spans nearly all the summer months when the heaviest amount of mist and rainfall can be expected.

Right: **Aloe nubigena** in its cliff-top habitat near Graskop in Mpumalanga where frequent mists condense and fall as soft, penetrating rain. Mosses, lichens and small tufts of grass often grow with the aloes in these elevated sites.

The aloes have been studied at three localities all near Graskop: God's Window with its towering moss-covered and lichen-blotched cliff faces, The Pinnacle, a well known landmark and featured in the illustration on page vi, and a gorge on the eastern outskirts of Graskop. The observations that follow are based on data collected at these localities.

The gorge on the eastern outskirts of Graskop

Some 30 years ago, the short dense grassland on the summit of the mountain above this gorge was well preserved, but the habitat has been progressively degraded over the years. Several developments have taken place immediately above the gorge, the grassland has been grazed by domestic stock and the boulder-strewn cliff-top habitat has been invaded by pines, which owe their origin to the nearby timber plantations.

Aloe nubigena used to be plentiful in the boulder-strewn, mossy habitat on top of the sheer south-facing cliffs at the eastern end of the gorge. The aloes grew in clumps amongst the boulders and on mossy banks. The plants have become scarce in recent times as much of the habitat has been smothered with pine needles. The seepage areas that used to irrigate the mossy banks have become degraded through overgrazing. The disturbed ground in the seepage areas has caused the proliferation of a large number of weeds, particularly Scotch thistles (*Cirsium vulgare*). *A. nubigena* is now confined to the limited amount of suitable habitat that remains—a small number of mossy ledges on the sheer cliffs and areas where a few plants have established themselves in rock crevices.

The cliffs around this gorge face south on one side and north on the other. *Aloe nubigena* is absent from the hot north-facing precipices which are in sun for the greater part of the day. The south-facing cliff faces where the aloes grow experience a few hours of direct sunlight in the early morning in summer. This habitat is in deep shade for most of the winter, at the time when the plants are dormant.

Habitat around The Pinnacle, just north of Graskop

The sheer cliffs near The Pinnacle contain a large number of mossy ledges and rock fissures, mostly around the uppermost section of the precipices. There is a limited amount of boulder-strewn cliff-top habitat, much of which is covered with dense indigenous bush. The lower quarter of the cliff faces also contain some broken ledges covered with lichens and moss. *Aloe nubigena* grows in these situations where the plants are irrigated from seepage above. In addition, a good deal of mist reaches them since they grow in the most elevated part of a gorge. The cliffs on the north-facing side of the gorge are in direct sunlight for much of the day. They are hot and dry and do not provide suitable habitat for the aloes.

Some clumps of moss in which *Aloe nubigena* grows dry out at the end of winter and detach from the cliff faces. This depletes the populations that grow in the thickest moss, which is most prone to falling from the cliffs. Most of the clumps of moss containing aloes fall into the forest below where the plants eventually succumb. Some, however, land on rock ledges where they take root. In this way these aloes are spread to places where they would otherwise be unlikely to occur.

The cliffs around God's Window

God's Window has some of the most spectacular escarpment scenery in the Graskop area. The cliffs reach great heights and contain much suitable habitat for *Aloe nubigena*. Most aloes grow near the summits of the precipices in compact banks of moss situated on narrow ledges. Others grow in cracks in the rock, containing little soil and often within the spray from seasonal waterfalls in the wet summer months.

The cliff-top habitat available to the aloes is limited owing to dense indigenous forest and large stands of *Clivia caulescens* tightly packed together. Most places suitable for supporting *Aloe nubigena* are occupied by these plants. The limited amount of habitat is one of the main factors restricting the distribution of these aloes across the cliff faces.

Growth cycle and the role of fire

Aloe nubigena grows from November to late March or early April, the time of the year when mists and rain are the most frequent. The plants have a long flowering period spanning all these months. Most clumps of the aloe come into bud and flower after a few days of dense mists or rain. Most plants flower well in late November and December, particularly if heavy mists and rain are interspersed with sunny weather.

The aloes are dormant during the dry winter and early spring, from May to September. They may produce leaves, however, after unseasonal heavy winter rainfall, which occasionally occurs in June or July.

During dry summer weather the aloes enter a state of semidormancy but resume growth and flowering once the habitat becomes moist again. Seeds are set throughout the later summer months. They are dispersed by wind and the forces of gravity, simply falling onto ledges below the parent plants. Very little seed lands in places where it is able to germinate. The limited release of seeds into the environment over a long time ensures that a few seeds land on suitable habitat where they can germinate and develop into young plants.

Periodic fires clear dead grass and ferns from the cliff tops. Much of this type of habitat around Graskop has been degraded by invasive alien trees and grazing livestock. Fires consequently have a very limited modern function in the little remaining grassy cliff-top habitat around Graskop. In addition, fires are actively prevented where there are plantations of alien timber trees.

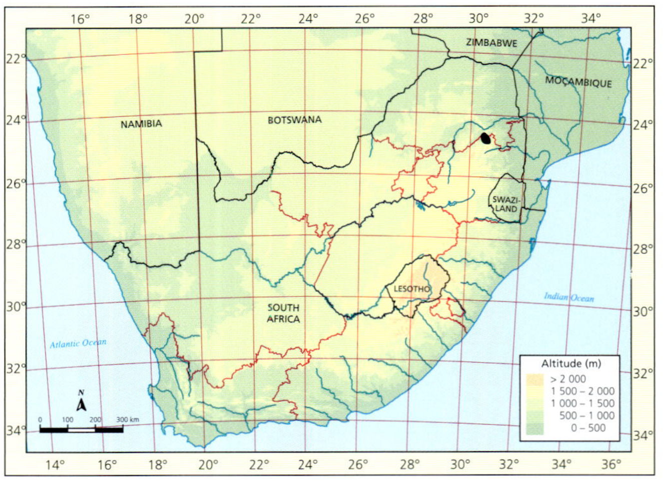

Distribution of *Aloe nubigena*

Status of wild populations

Aloe nubigena has disappeared from most of the grassy cliff summit areas where it used to occur around Graskop. Intact stretches of this part of the habitat are now rare and the aloes are mainly cliff dwellers. In the long term, the plants are secure in their rock face habitat. Degradation of seepage areas above cliff faces may affect some of the aloes by reducing the overall amount of moisture available to them. This is currently rather insignificant, owing to frequent dense summer mists. Should climatic change negatively affect the frequency of mists, degraded seepage areas are likely to cause some populations of *A. nubigena* to shrink or die out.

Aloe nubigena used to be frequent in some cliff-top areas near Graskop. The summits of precipices contain the most places suitable for the aloes. This fragile habitat is easily degraded and every effort should be made to protect these places where they still occur.

Cultivation, propagation and horticultural potential

This species is difficult to grow in cultivation. The plants require much regular moisture throughout summer when they grow and flower. Success is quite often achieved by cultivating them in large, moisture-retentive clay pots against a south-facing wall, which is in shade for most of the day in the southern hemisphere.

The aloes can also be grown in deep hanging baskets suspended from the branches of trees with dense foliage. They require filtered light, but a short spell of direct early morning sunlight is beneficial. *Aloe nubigena* grows well in an orchid house with mist spray and remains evergreen in this situation. To ensure the health of the aloes, they should be permitted to enter dormancy in winter. The container in which they are growing should be removed from the orchid house and kept largely dry in a shaded position for the duration of the winter.

Aloe nubigena suckers readily in cultivation and the species is easily propagated by removing some of the shoots and rooting them in moist sand. The stems should be snapped off the parent plant and left in a dry, shaded, well-ventilated position for a few days before planting. Cuttings are best rooted in spring, late summer or early autumn. Plants rooted in late summer or early autumn should be transplanted at the beginning of the following summer. *A. nubigena* thrives in an acid peaty soil. It grows well in a mixture consisting of equal parts of sieved peat, sandy soil or silt and sieved, very well-rotted leaf mould.

The aloes set some seed in cultivation if hand-pollinated. This should be lightly and evenly scattered across the surface of the same soil mixture used for cultivating adult plants. The seeds should be very lightly covered with soil and allowed to germinate in places with dappled shade. Deep seed trays or pots are equally suitable for germinating seeds.

Aloe nubigena, like *A. soutpansbergensis* and *A. thompsoniae*, has good potential as a subject for hanging baskets in general horticulture.

21. *Aloe parviflora*

Aloe parviflora Baker

Aloe parviflora has an interesting history. It was described by Baker (1901: 785) from plants collected in Pinetown on the outskirts of Durban in KwaZulu-Natal. Specimens were not located again until many years later, and not at the time Reynolds was engaged in researching and writing his magnum opus *The aloes of South Africa* (Reynolds 1969: 113–115). Jeppe (1996: 286) took the view that *A. parviflora* may only represent a regional form of *A. minima*. Glen & Hardy (2000: 9) reduced *A. parviflora* to synonymy under *A. minima* without supplying any reasons for their decision. It is a pity that this species was not relocated during Reynolds's fieldwork. No comparable field research has been initiated on South Africa's aloes since that time and it would have been useful to have had Reynolds's opinion on the correct placement of *A. parviflora*.

In this work, *Aloe parviflora* is accepted as a good species for various reasons. As Glen & Hardy (2000) do not account for their reduction of *A. parviflora* to synonymy, there is an anomaly in the knowledge construction process. Knowledge can be advanced only by supplying reasons for decisions made when species are reduced to synonymy. These reasons can then be analysed and critically reviewed by future workers and form part of the process used to develop new species concepts in any genus. In the absence of such reasons, the status of knowledge reverts to the previous paradigm.

The species concept of *Aloe parviflora* has to be understood with reference to detailed sampling across *A. minima* and *A. parviflora* populations if they can be located in the same or adjacent areas. Should the two species be easily identified where they occur sympatrically, an account has to be provided of the processes that keep them separate. On the other hand, if the two plants intergrade, reasons have to be provided for the circumstances that permit this occurrence.

Aloe minima and *A. parviflora* have a peak flowering time during the first three weeks of February and as a result different flowering seasons could not maintain them as separate species where they occur together. The area most likely to support both species lies in the hilly area between Durban and Pietermaritzburg. This region is densely settled and developed and much of the remaining grassland has

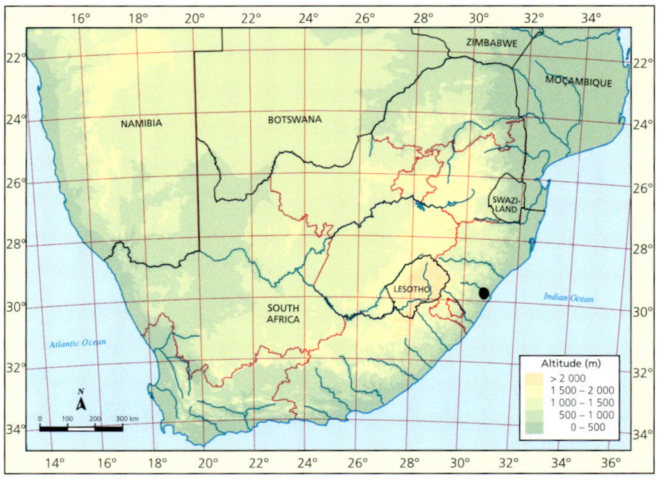

Distribution of *Aloe parviflora*

been degraded and transformed by invasive alien plants. The remaining habitat is probably too fragmented for any meaningful conclusions to be drawn, even if *A. minima* and *A. parviflora* can be found growing sympatrically.

Aloe parviflora is now known mainly from the Botha's Hill and Drummond areas not far from Pinetown where the species was originally located. The aloes are also found in the little remaining grassland habitat south of Camperdown and in The Valley of a Thousand Hills, two areas adjoining Botha's Hill and Drummond. The species has been studied in detail near Botha's Hill for a period of about 10 years and these observations provide the details for the account that follows.

Near Botha's Hill

The study area near Botha's Hill consists of a range of hills surrounded by suburbs at the southern end. The habitat boundaries to the north lie between suburbia and the dense rural settlements of The Valley of a Thousand Hills. The study site consists of a narrow strip of grassland only about 400 m wide at its broadest point. This limited area does, however, contain the relatively many places occupied by this species. The ability of *Aloe parviflora* to colonise many places is unusual for a grass aloe species and it therefore occurs commonly in the limited amount of remaining habitat.

The plants are found in shallow soil over sheets of exposed sloping sandstone, in mats of *Selaginella* or clumps of moss. Here they share their habitat with *Brachystelma pulchellum*, a caudiciform commonly found in these situations. They also occur in rocky places with thin soil and minimal grass cover in southeast- and north-facing positions. It is unusual to find such a small grass aloe on hot northern hill crests. The aloes are common amongst short but dense grass in level or gently sloping areas on the summits of the hills. They particularly favour the upper southwest-facing hill slopes where the grass cover is dense. They even occur in shallow sand on rocky ledges at the top of a road cutting truncating the side of a hill.

The *Aloe parviflora* habitat near Botha's Hill represents a microcosm of changes to the grassland bordering settled areas in eastern South Africa. The area nearest to The Valley of a Thousand Hills is steadily being degraded by grazing livestock, particularly cattle and goats. Most of the aloes growing in banks of moss on sandy soil over rock sheets have been trampled and destroyed. The disturbed soil has also attracted annual and perennial weeds, which are in the process of degrading the habitat even further. The aloes are now confined mainly to rocky habitats where they escape trampling. However, they still occur in grassy areas amongst rocks, which are the least attractive to grazing livestock.

The remainder of the habitat hemmed in by suburbia is well preserved. The aloes are common here in all habitats usually occupied by this species. The grassland is steadily being invaded by alien vegetation, particularly bugweed (*Solanum mauritianum*).

Aloe parviflora flowers well each year, even during dry weather. There is normally an abundant seed set except in years when seed wasps are abundant. Seedling plants are well represented in the habitat and urban development has made little, if any, difference to the pollination biology of the plants.

Growth cycle and the role of fire

Aloe parviflora is a summer-growing species. The plants start to sprout leaves in November, a time when the first good rains and mists of summer usually occur. Growth continues during December and January. Flower buds start to form in the last two weeks of January and the plants come into flower in the first two weeks of February. Few plants flower before or after the main flowering period. The numerous buff-coloured, capitate racemes are conspicuous when the aloes are in flower. Simultaneous flowering is likely to play a role in attracting large numbers of pollinators as indicated by the abundant seed set prevalent from one season to the next.

Most of the aloes fail to come into flower during prolonged summer droughts. The exceptions are plants growing in places where they frequent shallow soil over sheets of exposed sandstone or south- or southwest-facing hill slopes. These sites retain some moisture after dense mists or light rain and provide sufficient moisture to bring the plants into flower.

Seeds are dispersed by wind and water runoff after rain. They are mostly liberated into the habitat during the last three weeks of March at a time when rainfall is decreasing and some plants are already starting to enter dormancy. Most seeds lie dormant until the following summer rains begin when they have their first opportunity to germinate.

Fire is necessary for clearing the habitat from dead grasses. The grass cover is only thick enough to burn thoroughly once every two to three years and this attribute of the grassland favours the development of seedling plants. Flowering tapers off if the veld remains unburnt for three years or more, except for plants growing in moss over sheets of rock. Burning of the habitat is becoming progressively more uneven around Botha's Hill owing to increasing invasion of the veld by introduced plants.

Status of wild populations

The status of *Aloe parviflora* depends to a large extent on whether it would ever be feasible under modern circumstances or legitimate to reduce it to synonymy under *A. minima*. If it is, then *A. parviflora* is only a regional form of *A. minima* and could be described as a subspecies or variety. If, on the other hand, the species concept *A. parviflora* stands the test of time and innovations to the botanical knowledge production process, then it may well be classified as vulnerable or endangered.

In the Botha's Hill, Drummond and Cato Ridge areas, *Aloe parviflora* may well disappear as the years progress unless there is some attempt to conserve the remaining grassland where they occur. Habitat would have to be protected from further suburban development and kept free from invasive alien plants. It is significant in this respect that the biological processes of fire and pollination, which sustain the populations of this aloe, have been little affected by suburban developments.

Right: ***Aloe parviflora*** growing in short grass amongst rocks near Drummond in KwaZulu-Natal. This diminutive species is still well represented in a few patches of rocky grassland within suburbia.

Cultivation, propagation and horticultural potential

Aloe parviflora is one of the few small grass aloes that are easy to cultivate and set seed readily in cultivation with the aid of hand pollination. The aloes should be grown in sharply drained soil, preferably in earthenware containers. The ideal cultivation mixture consists of half sieved sandy soil to which equal amounts of silt and very well-rotted leaf mould or kraal manure have been added. The components of this cultivation medium should be very thoroughly mixed together.

Plants should be grown in morning sunlight or light shade and watered at intervals throughout the summer, until night temperatures drop at the beginning of autumn. The leaves usually die back completely in April and early May in the southern hemisphere. The aloes should be lightly and occasionally watered in winter and more regularly in late spring, just before the start of the new growing season.

Seeds should be sown in late spring and early summer to give the seedlings enough time to develop before the winter dormancy period. The seeds should be lightly and evenly scattered across the surface of the soil. The same mixture in which the adult plants are grown can be used for germinating seeds. Seedlings should be well watered but care has to be taken that they are not kept too damp. Until they have developed their first set of leaves, young plants are particularly prone to attack by damping-off fungi. Seedlings grow and develop as well in large, deep seed trays as they do in deeper pots. The young aloes are usually ready to transplant at the end of their second or third growing season.

22. *Aloe saundersiae*

Aloe saundersiae (Reynolds) Reynolds

This is the smallest grass aloe and one of only a few species that grow in clumps of moss on rocky outcrops as well as in short rocky grassland. It is found in rocky, mountainous country near Nkandla above the headwaters of several minor tributaries of the Tugela River. The Nkandla area of KwaZulu-Natal is densely settled by pastoralists and small-scale stock farmers. The number of rural homesteads has expanded over the years and many of them are close to *Aloe saundersiae* habitat. Plants in open grassland are trampled by grazing livestock and the species has disappeared from tracts of mountainous grassland where it was once numerous.

Colonies southeast of Nkandla in central KwaZulu-Natal

Aloe saundersiae has been studied in the mountainous area southeast of Nkandla in KwaZulu-Natal for a period of nearly two decades. The plants grow at an altitude of 1 000–1 300 m in rocky, mountainous grassland southeast of the town of Nkandla. This area is in the vicinity of the type locality and the study has provided an opportunity for assessing changes to the habitat brought about by increased settlement, increases in livestock numbers, erosion and too frequent burning of the veld.

The species is found in three distinct types of habitat on the mountains—patches of short grassland on steep slopes below rocky outcrops and cliffs, grassy patches on and beside sheets of exposed rock and banks of moss on low cliffs or rocky outcrops. The plants were by far the most numerous on steep, grassy slopes below cliffs. This habitat is rather limited and sometimes only about 50 m wide. It usually marks the perimeter of indigenous forests below broken south- and southeast-facing mountain summit cliff faces. When the area was first visited, there were large populations of *Aloe saundersiae* amongst the grass on the steep hillsides. The larger populations of plants contained seedlings of various ages and some plants were also growing on the cliff faces in clumps of grass and moss on the wider ledges.

In former times, settlements were confined to the level or greatly sloping mountain summit grassland. More recently homesteads have been founded close to *Aloe saundersiae* habitat, resulting in extensive grazing of the grassy slopes by goats, trampling of the plants and alteration of the habitat through erosion. The eroded habitat has been colonised by weeds and *A. saundersiae* now occurs only on wide cliff ledges amongst moss or grass.

The southeast- and south-facing mountain slopes consist of short, rocky grassland periodically saturated during misty or rainy weather. *Aloe saundersiae* is widely scattered across this habitat, occurring either singly or in small clumps of about two to six plants. This habitat dries out

Right: ***Aloe saundersiae*** in flower amongst mosses and small, short grass tufts on a rocky outcrop near Nkandla in KwaZulu-Natal. Rocky outcrops represent the only remaining habitat where these small aloes are able to survive.

thoroughly during sunny weather, except for the seepage areas which remain moist for long periods in summer. The aloes tend to be more numerous in the vicinity of seepage areas where the seed is afforded the best conditions for germination. Seedlings also have enough moisture to develop into young plants, a situation that is not found further away from the seepages.

Most of the habitat consists of short grass tufts and sandy patches between rocks, with a variety of small bulbs as well as tuberous plants such as *Brachystelma modestum*. The same habitat is found on the hotter north- and north-west-facing mountain slopes but *Aloe saundersiae* is mostly absent here. The habitat becomes too dry for these diminutive aloes to survive.

A number of the rocky outcrops on the mountain summits harbour small cliffs with mossy ledges. *Aloe saundersiae* grows in small groups amongst the moss, sometimes only the upper portions of the leaves protruding above the moss. The species is infrequent in clumps of moss and is currently found mostly in rocky grassland. This habitat is unattractive on the whole to grazing animals. Although the plants are little affected by grazing animals, they are thinly and widely scattered across the habitat. The main reason is that, apart from the vicinity of the seepage areas, the growing conditions are harsh and the soils very thin.

Growth cycle and the role of fire near Nkandla

Aloe saundersiae is a summer-growing plant with its peak flowering season in the first two weeks of February. The habitat is subject to prolonged and frequent mists during the height of summer, from November to February. The mists augment the already high rainfall. The leaves emerge from September to November, depending on the arrival of the main summer rains after the cold, dry winter. The peak of leaf production is usually November and early December.

Flower buds start to form at the base of the leaf rosettes in early January. Dry, hot summers result in poor flowering, with many plants in more arid sections of the habitat failing to come into flower. The ideal flowering conditions are provided by a wet early summer followed by dry sunny days in early February. Pollination is very poor in years when the weather is misty and rainy during the flowering season.

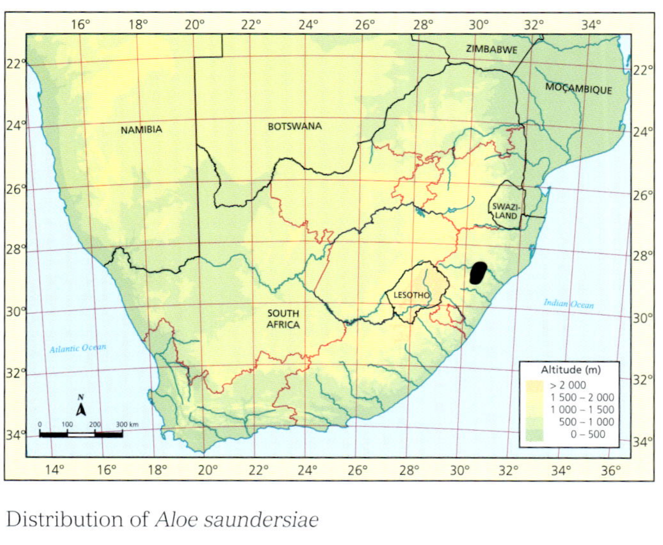

Distribution of *Aloe saundersiae*

Fire is the most significant agent for keeping the habitat clear from accumulations of dead grass. The grassy slopes on which the aloes used to occur were regularly burnt in former times but now burn only superficially owing to the degraded nature of the habitat. Fires periodically burn clumps of dry moss in which the aloes are growing but the moss is only rarely dry enough and long enough to burn thoroughly. Moderate grazing, such as that which occurred in times gone by, and periodic fires helped to keep the habitat clear of dead grasses and herbs. Fires are now more frequent than in the past owing to denser human settlement. Furthermore, they now sometimes occur in summer during dry weather, thereby destroying flowers and ripening seeds of *Aloe saundersiae*.

Pollination of flowers in dense groups of *Aloe saundersiae* growing on the hill slopes used to be good in former times. The many stems of flowers in close proximity to one another were attractive to pollinators, resulting in a good seed set. Pollination of the plants on the rocky outcrops close to the large groups of aloes was also good. Fewer flowers borne on plants growing on the outcrops are pollinated under the changed environmental conditions. The large groups of plants certainly functioned to attract pollinators to the general habitat in former times.

Aloe saundersiae liberates its seed into the environment in April towards the middle of autumn. This is a time when mists and rainfall decrease sharply and the habitat begins to dry out. The seeds are dispersed by wind and water runoff from the first summer rains the following season.

Status of wild populations

Populations of *Aloe saundersiae* were healthy near Nkandla until recently, occupying all the suitable habitats at this location. As a result of too frequent burning, grazing pressures and soil erosion, the plants have declined sharply in numbers. The optimum habitat consisting of steep grassy banks below cliffs has now largely been degraded and taken over by weeds and other pioneer vegetation. The species is now confined to the least suitable habitat and may disappear from much of this as well, particularly under conditions of global warming. The example of the plants growing near Nkandla shows how rapidly changes to the habitat can affect the status of the smaller grass aloes.

Reynolds noted (1969: 111) that *Aloe saundersiae* was recorded from the Lowlands area, at Harden Heights and in the hilly area near the Umvoti River between Pietermaritzburg and Greytown. Most of these areas of KwaZulu-Natal have been heavily afforested with alien timber trees or densely settled by rural pastoralists. The status of the plants in these regions has to be investigated to form a yardstick for comparison with the Nkandla populations.

Cultivation, propagation and horticultural potential

Aloe saundersiae is an attractive species suited to cultivation in containers. An effort has to be made to ensure its long-term survival in cultivation since the plants are prone to fungal attack, root mealybug and snails which eat the leaves and predispose the plants to rot in rainy or humid weather.

The plants should be grown in coarse, granular sand to which some very well-rotted, finely sieved compost has been added. Another good mixture consists of one third sieved sandy soil, one third coarse, granular sand and one third very well-rotted, sieved compost made either from peat moss or veld grass.

The plants have to be well watered throughout the summer growing season. Earthenware containers are more suited to the cultivation of this species than plastic pots. They prevent overheating of the soil in the midsummer months which predisposes the aloes to fungal attacks and rot. This species should be grown in lightly dappled shade or morning sunlight. The containers may be watered lightly and occasionally in winter, otherwise they should be kept dry.

These aloes are best propagated from seed. Seeds should be lightly and evenly scattered across the surface of the soil and only just covered with the planting mixture. They should be well watered and kept in light or dappled shade with plenty of free air movement. Seeds are best sown in the spring or early summer, affording the tiny seedlings the chance to establish themselves before the winter. The seedlings, like the adult plants, should be kept dry in the winter and watered only occasionally and sparingly. Seed trays should contain a large amount of drainage material at the bottom forming a base layer of 1–1.5 cm. Seed is sparingly produced in cultivation and plants usually do not set seed without the aid of hand pollination.

Aloe saundersiae is often difficult to maintain in cultivation in the long term and as a result may never become a popular cultivation subject. Seed is rarely available, but this tiny grass aloe may be suited to multiplication via the techniques of tissue culture.

23. *Aloe soutpansbergensis*

Aloe soutpansbergensis I.Verd.

Aloe soutpansbergensis is confined to cliff faces and the highest peaks of the central and western sections of the Soutpansberg, an isolated mountain range in the northern part of South Africa's Limpopo Province. The plants grow in equivalent habitat to that of *A. nubigena* on the escarpment cliffs near Graskop and the summit peaks and vast cliffs of the Wolkberg inhabited by *A. thompsoniae*. All these species are mist-belt plants obtaining much of their moisture from light but penetrating rain that falls during dense mists.

The summits of the central and particularly the western parts of the Soutpansberg contain several high peaks, mostly at an altitude of 1 600–1 700 m. Mists swirl up from the southern side of the mountain range and these peaks receive the greatest amount of moisture. The peaks on the northern side of the mountain, even though they are almost as high, tend to receive much less mist. This habitat is mostly too arid for *Aloe soutpansbergensis*. The aloes are common where they occur, particularly if they can utilise the full range of its preferred habitat as at Letjuma, the highest peak in the western Soutpansberg. Much of the habitat on the mountain range is unsuitable for the plants. This applies to virtually all the northern slopes of the mountain.

Aloe soutpansbergensis was described by Verdoorn in 1962. This interesting description and account of the plant in its habitat provides important details as to the knowledge production process concerning the description of new aloes at the time. It was debated for a while whether *A. soutpansbergensis* was distinct from *A. verecunda* or merely a variety of that species. The matters discussed focused on floral and foliage differences. The argument for the distinctiveness of this species could, however, have been strengthened with reference to specific adaptations these aloes have for growing in shallow soil on rocks in a mist-belt climate. These include a fine, laterally spreading root system well adapted to absorbing moisture provided by mists. *A. verecunda* and *A. vossii* have extensive, thickened root systems which do not predispose them to habitats with the thinnest of soils such as those occupied by *A. soutpansbergensis*.

John Lavranos was by chance preparing a description of *Aloe soutpansbergensis* at about the same time when the

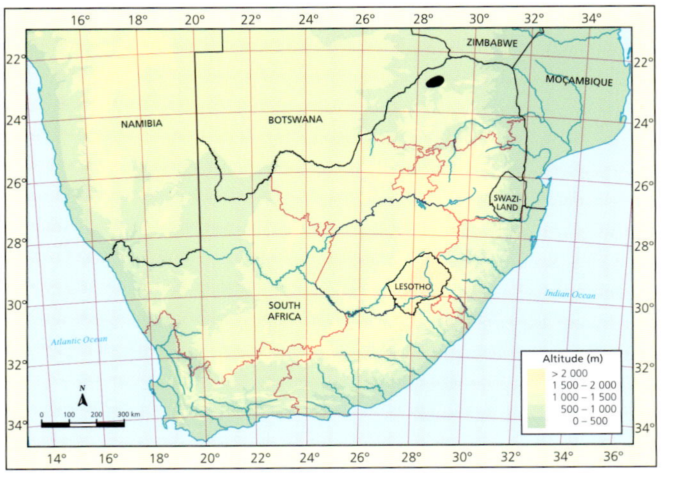

Distribution of *Aloe soutpansbergensis*

Letjuma, the highest peak in the Soutpansberg, the most northerly mountain range in South Africa. The rugged rocky outcrops, mossy cliff faces and short grassland provide extensive habitat for *Aloe soutpansbergensis*. Letjuma is the type locality of this dwarf *Aloe* species. The higher peaks in the Soutpansberg are frequently bathed in dense mist during the rainy summer months, with a periodic covering of light, misty cloud in winter. The mountain peak was drawn from a grassy vlei at the foot of the summit.

species was painted for description in *The Flowering Plants of Africa*. He sent his draft manuscript to Dr Verdoorn after learning that she was in the process of describing the species. John specifically referred to the fact that *A. soutpansbergensis* was adapted to mist-belt conditions on mountain peaks. The Soutpansberg experiences prolonged droughts and during this time the small stature of the plants and their fine root system ensures that they survive on moisture provided by mists.

Aloe soutpansbergensis has been studied at two localities on the central and western Soutpansberg: the sheer summits of the cliffs above Schoemansdal in the central Soutpansberg and Letjuma, the highest peak in the western Soutpansberg.

At Letjuma

Letjuma is a magnificent, broad mountain peak in the western Soutpansberg surrounded by patches of grassland with indigenous forest below. The habitat consists of tumbled boulders, small cliffs and large sheets of steeply sloping, lichen-blotched rocks. The aloes occur on these steeply sloping rocks where they grow in shallow soil amongst tufts of short grass and moss. They are also particularly numerous along the upper edges of small cliffs. These sites are well irrigated from seepage areas above. In these situations the aloes grow amongst moss and short grass. Other well-utilised situations are shallow pockets of soil at the edges of sheets of exposed rocks and fissures under large rocks and boulders.

The plants avoid areas where the grass is dense or tall and areas covered in bracken (*Pteridium aquilinum*). Bracken grows mostly on the lower slopes together with taller grasses. *Aloe vogtsii* is abundant on these lower slopes, but much less plentiful in the areas of thinner soil where *A. soutpansbergensis* thrives. In situations where *A. soutpansbergensis* remains moist for longest after mist or rain, it suckers profusely and large mats of this species are frequently encountered.

The habitat becomes very dry at the end of winter, in July and August. Clumps of moss and grass break off the boulders at this time of the year and if they contain stems of the aloes, the plants are distributed further in the rocky habitat. Some moisture is provided by heavy dew and light fog patches in the winter months.

On the cliffs above Schoemansdal

The Soutpansberg above Schoemansdal is characterised by tall, sheer cliffs. The upper reaches of these cliffs consist of broken rock faces with tufts of short grass and moss. This restricted habitat is limited to the upper quarter of the cliff faces. The area receives most of its precipitation during mists, even if they are light and of short duration, which is often the case during droughts. The aloes occur sparsely under these situations, sometimes even growing amongst dense lichens where pockets of humus have been trapped in small spaces.

The species is absent from much of apparently suitable habitat on the rocky summit of the mountain above the cliffs. These situations are at a slightly higher altitude but there is not as much precipitation from mists as there is on the upper reaches of the cliff faces.

Growth cycle and the role of fire

Aloe soutpansbergensis is a summer-growing species that produces foliage and flowers from November to March, the time when there is the most precipitation. The plants usually flower after dense mists lasting a few days, or after heavy rainfall.

The aloes set seed throughout their growth cycle, but little is produced during dry summers as the plants flower very sparingly. Seeds are dispersed by wind, which is rarely absent from the cliffs and mountain peaks in the Soutpansberg in summer. A large proportion of the seed

Right: ***Aloe soutpansbergensis*** in its cliff-top habitat on the Soutpansberg in the northern corner of Limpopo Province. The aloes typically grow amongst rocks and short grass tufts and are still relatively common as their habitat lies mostly to the west of the main timber-producing area, well preserved in private nature reserves.

J Condy

goes to waste owing to the rocky nature of the habitat. Despite this, many places that can be occupied by this species are filled to capacity by clumps of these aloes, as at Letjuma. The summit of the Soutpansberg starts to dry out towards the end of March and the plants are dormant from April until the major summer rains begin late the following October or in November.

Aloe soutpansbergensis flowers and grows well in the absence of fire where it occurs in lichen- and moss-covered areas virtually devoid of grass. Most of its habitat, however, contains some grass which may become dense in years between fires. Fires clear the habitat of moribund grass and are also responsible for redistributing some nutrients to the patches of thin soil frequented by *A. soutpansbergensis*.

Status of wild populations

The species is usually common where it occurs, but the habitat suitable for the plants is limited to places with large amounts of mist during the summer months. These cliff-dwelling aloes are subject to few pressures and their populations are limited in size by the scarcity of habitat.

The plants are most common on mountain peaks that offer the full range of habitat conditions, these days falling largely within the boundaries of extensive privately owned nature reserves. This form of land use is conducive to the long-term conservation of the species. The reverse would be true if areas of the central and western Soutpansberg were to be used for indigenous pastoralism. Goats would quickly trample many of the plants growing in exposed positions with thin soil. The area around Letjuma in the western Soutpansberg has extensive grasslands and has been used for cattle ranching in the past. This has made little difference to *Aloe soutpansbergensis* since there is insufficient grass on the mountain peak to attract grazing cattle. The greatest danger to *A. soutpansbergensis* may lie with the consequences of global warming, particularly if it affects the frequency and duration of mists along the mountain range.

Aloe soutpansbergensis has never been comprehensively surveyed on the Soutpansberg. It would be useful if the ratio of cliff-top to mountain-peak populations could be established since the cliff-growing plants are largely exempt from any changes to current land use practices. There are several difficulties associated with surveying populations of *A. soutpansbergensis*—the plants grow in remote and inaccessible parts of the mountain range and their distribution across the mountain is uneven.

Cultivation, propagation and horticultural potential

This is one of the grass aloe species well suited to cultivation. It grows particularly well in hanging baskets filled with peaty soil and kept moist throughout the summer months. The baskets can be hung from the branches of trees that produce light shade or from trellis work. The plants require plenty of free air circulation but should be kept out of strong direct sunlight, except in coastal regions with a mild climate. The plants should be watered occasionally during spells of warm weather in winter. The frequency of watering should be increased during September and October in preparation for the beginning of the growing and flowering period.

This species proliferates from stem cuttings and can be multiplied rapidly via this method of propagation. The stems can be rooted either in coarse sand or directly in gritty, peaty soil in hanging baskets. The aloes do not grow nearly as well in pots as they do in hanging baskets. The soil in pots tends to compact too much after a few seasons unless the cultivation mixture is very porous. Porous mixtures, however, often render the plants susceptible to attacks by root mealybug.

Aloe soutpansbergensis can be grown from seed but this is hardly necessary in view of the ease with which it can be propagated from suckers or stem cuttings. Flowers should be hand-pollinated with a fine artist's brush to improve seed yield. Seeds should be evenly sown across the surface of a soil mixture consisting of one part fine and one part sieved peat or leaf mould that has been left out in the sun to dry for a few days. The pots used for sowing should be regularly watered until the seeds have germinated and the young plants have developed. Watering should be gradually decreased in the autumn and early winter months. The young aloes should be kept largely dry during their first winter dormancy but may be lightly watered during periods of mild weather.

Aloe soutpansbergensis is a suitable candidate for general introduction to horticulture as it is easy to grow and propagate, compact in habit and very floriferous for much of the summer under cultivated conditions.

24. *Aloe thompsoniae*

Aloe thompsoniae Groenew.

Aloe thompsoniae is another species confined to a mist-belt mountaintop habitat. The plants grow on the summit of the Wolkberg, the northernmost section of the Drakensberg, in Limpopo Province. The species was discovered by Awdry Thompson after whom the plant was named. Awdry and her daughter Sheila made several significant contributions to the study of the flora of the Wolkberg area. A general account of this subject is given in the book *Between Woodbush and Wolkberg* (Wongtschowski 1990: 171–182).

The species has been little affected by the extensive afforestation around the Wolkberg, all of which is well below the mountain summit. The plants are similar to *Aloe nubigena*, a species found in mountaintop mist-belt habitat near Graskop. There are, however, some significant differences between the autecology of the two plants. *A. thompsoniae* occupies three distinct habitats on the Wolkberg. The plants grow on sheer precipices around the edges of large boulders, on smaller rocks and in seepage areas over extensive sheets of exposed sloping rock with very short grass and moss cover, high up on the summit of the mountain. *A. nubigena* grows mostly on mossy cliff faces in its habitat near Graskop and much more rarely on moss-covered boulders around the summit of these cliffs.

Aloe thompsoniae is able to take advantage of a wide range of habitat on the summit of the Wolkberg, all with different levels of summer moisture. This accounts for the fact that the plants are numerous in a limited area. In addition, the suckering habit of this species is strongly developed. The plants are frequently encountered in large clumps with many hundreds of stems.

The time chosen for the field research on the Wolkberg was 21 December 2003, just before the longest day of the year in the southern hemisphere. Roger and Wilna Stones, two keen amateur botanists from Haenertsburg, had located several plants in flower the week before. They kindly accompanied the author and artist to the *Aloe thompsoniae* type locality to conduct field research and paint the plants in habitat. Some rain had fallen on the Wolkberg before the visit and a veld fire had swept across the mountain the previous winter. These conditions were ideal for research.

On the summit of the Wolkberg

The sites occupied by *Aloe thompsoniae* allow it to take advantage of the widest range of exposed rocky habitat periodically moistened by mist and rain on the Wolkberg. The plants are not usually found amongst taller grasses and low shrubs, nor in dense grassland away from rock outcrops. Those growing on sheer cliff faces are found in south-facing positions where they are in shade for much of the day. They also occur on northern cliff faces, in situations where they are well irrigated by seepage from above. They tend to be more abundant on southern cliff faces as these situations retain moisture for longest after mist or rain.

The aloes growing on sheer precipices are usually the first to flower after the beginning of the main summer rains. Plants in more exposed positions, on rocky outcrops, may only be in the initial stages of bud formation when those growing on rock faces are in full flower. Cliff-face populations expand and regenerate extremely slowly from wind-blown seed that lands in protected places where it

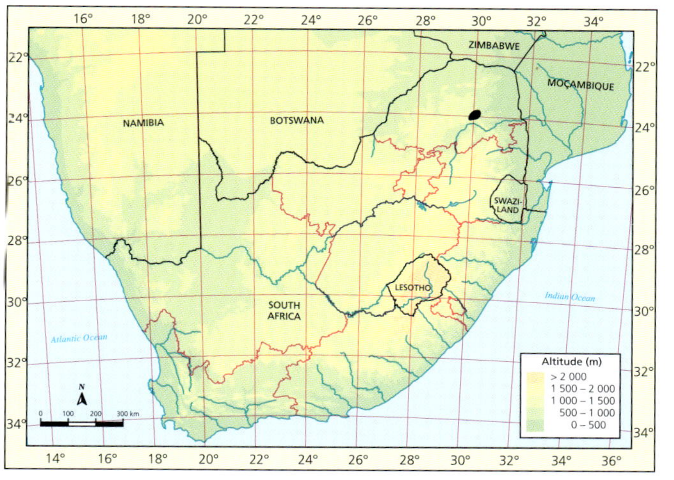

Distribution of *Aloe thompsoniae*

can germinate and take root under misty and rainy conditions. Young single-stemmed plants, only a few centimetres tall and growing near the summits of the cliffs, provide evidence of this process.

Aloe thompsoniae is very abundant in short grass and short moss over sheets of exposed rock. These areas funnel off water after dense mists and rain. They are also periodically very wet and very dry, mostly exposed to direct sunlight for much of the day. These situations remain moist for much of the summer during years of good rainfall. During droughts even small amounts of rainfall and mist provide the plants with enough moisture to produce leaves and some flowers. This was evident during fieldwork on 21 December 2003. The preceding months had been very dry, but some of the plants on the driest exposed rock sheets had produced leaves and a few had started to initiate flower buds.

Much of the seed produced by plants growing on rock sheets germinates well. Most, however, lands within large clumps of aloe stems where the resultant seedlings make little difference to the overall population size of the plants. Some seeds, however, are blown by wind or washed by rain into areas where there are no existing groups of aloes. Small single-stemmed or diminutive clumps of *Aloe thompsoniae* are an indication of this process.

The aloes are also commonly found at the edges of boulders or rocks, the most abundant habitat on the mountain summit. These populations are widely scattered over a large area. Each group of closely packed stems is generally small, limited by thick growth of tall moss, grasses, low shrubs and masses of closely packed bulbs, probably members of the *Merwilla plumbea* complex. The aloes also compete with *Ledebouria* bulbs, which densely colonise this specific habitat.

Growth cycle and the role of fire

Aloe thompsoniae grows and flowers in summer. The plants produce leaves with the onset of the main summer rains in November. If rains and mists are late in the season, the growing and flowering period will be delayed accordingly. The aloes usually come into flower in December and those growing on cliff faces, which retain the highest amount of moisture, are usually in full flower at about the time of the southern summer equinox.

Seeds are liberated into the environment in February and March. If the late summer and autumn months are misty and rainy, some of the seeds germinate then, or otherwise the following summer.

Fire plays a significant role in keeping the habitat clear of moribund grass and accumulations of dead herbage. Most fires occur in the winter months, from May to September, while the aloes are fully dormant. Fire promotes the flowering of *Aloe thompsoniae*. Many plants were seen in bud in December 2003 after the habitat had been thoroughly burnt during a grass fire the previous winter. It is probable that clearing of the habitat, rather than the effects of burning itself, promotes flowering since plants growing on cliff faces flower well despite the absence of fire. Cliff-face habitats have very little, if any, grass and the little cover that exists is generally very short. Fire redistributes nutrients in the habitat, a situation commonly found in localities inhabited by most grass aloe species.

Status of wild populations

Aloe thompsoniae is well distributed and common on the Wolkberg. This mountaintop habitat forms part of the Wolkberg Wilderness Area, which affords the plants formal protection.

The timber industry has no effect on the plants, unlike nearly all other grass aloe species found in South Africa. Stray pine and wattle seedlings that take root on the summit of the Wolkberg are usually removed when they are noticed. *Aloe thompsoniae* represents one of the rare cases in which neither the actual habitat nor its hydrology has been affected by afforestation. Timber plantations and clumps of alien trees growing below the Wolkberg were being felled at the time of the fieldwork on *A. thompsoniae*. The removal of alien vegetation, if followed up to remove seedlings, is likely to permit the rehabilitation of the grassland in the coming decades.

Cultivation, propagation and horticultural potential

Aloe thompsoniae is easily cultivated in a well-drained peaty soil consisting of one third finely sieved peat or acid, well-rotted compost, one third coarse-grained sandy soil and one third river sand. The plants should be grown in pots kept in morning sunlight or very lightly dappled shade. The aloes will grow in direct sunlight but then rarely flower well and require extensive watering during the summer growing period. Plants grown in wide, deep containers sucker freely and produce impressive clumps with a succession of flowers during the midsummer months.

Propagation of the plants is carried out by splitting clumps into single stems and spacing these about 6–8 cm apart in deep propagation trays. These stems produce a cluster of roots after about a month in the summer and they also send out suckers from the base. At this time the young plants can be potted up and further cuttings can be taken when the clumps of side shoots are about 4 cm tall. This species is particularly attractive when grown in rough-textured terracotta pots that attract moss to their surfaces. It may also prove possible to mass produce this grass aloe through tissue culture, a process successfully carried out with other smaller grass aloe species.

Propagation from seed hardly seems merited since the plants are so readily produced from cuttings and offsets. Seeds should be lightly and evenly scattered across the surface of the soil and just covered. The soil mixture used for the cultivation of plants in pots and striking cuttings is also ideal for the germination of seeds. The sowing mixture should be kept moist throughout the germination period and also while the seedlings are developing. Care has to be taken that seedlings do not damp off. A copper-based fungicide is usually effective for preventing damp off and may have to be used repeatedly during damp or humid summer weather. Cuttings should be rooted in spring, summer and autumn and seeds sown in spring and summer.

Aloe thompsoniae, in common with all South African grass aloe species, requires a dry winter dormancy period. Plants should be lightly watered occasionally during periods when winter temperatures are mild.

Right: ***Aloe thompsoniae*** growing on a cliff face with an eastern aspect in rugged mountain summit habitat on the top of the Wolkberg ('Cloud Mountain') in Limpopo Province. The aloes grow amongst short grass tufts and moss, frequently swathed in mist during the summer months.

25. *Aloe verecunda*

Aloe verecunda Pole Evans

Aloe verecunda was described from plants collected in January 1913 at Ashmoledales in the Wolkberg, Limpopo Province. The species has come to be known, however, from plants formerly found abundantly on the rocky ridges around Johannesburg and southeast of Krugersdorp in Gauteng. These days the species appears to be best represented on the rocky ridges of western Gauteng and the adjacent part of the North-West Province. This area falls within a region with large numbers of deep-level gold mines. Gold mining has inadvertently preserved much of the habitat and some very large colonies of *A. verecunda* are to be found near Carletonville and Westonaria. The species is still found in the hills near Krugersdorp and Melville, growing on the few rocky areas that have not been incorporated into suburban development.

It is also found near Middelburg in Mpumalanga and in the Athalia area east of Piet Retief, also in Mpumalanga, close to the Swaziland border. Some of the plants east of Piet Retief are difficult to distinguish from plants growing in *Aloe vossii* populations on the upper northern slopes of the central Soutpansberg.

Aloe verecunda is very variable across its large geographical range. It is apparent from the data collected for this book that the currently understood species concept may have to be redefined. Afterwards clarity will be obtained as to the correct placement of *A. fouriei* and *A. vossii*.

Some interesting examples of *Aloe verecunda* that could be used towards defining the theoretical boundaries between *in situ* and *ex situ* conservation were examined on 29 December 2004. These plants were growing in a garden in the suburb of Breaunanda in Roodepoort near Johannesburg. Breaunanda and the adjacent suburbs once formed part of an extensive area of rocky ridges where *A. verecunda* was commonly found. The garden, which belonged to Mr Paul Asquith, had been preserved as a piece of natural rocky grassland and was periodically burnt during the dry winter months. There were several clumps of *A. verecunda* growing on the property and also others on a neighbouring rocky outcrop too precipitous for suburban development.

The aloes had flowered well owing to a grass fire earlier in the season. Many flowers had been pollinated and were starting to form seed capsules. These aloes were certainly preserved *in situ*, having been part of a much larger population before suburbia was developed. The plants are, however, technically speaking growing in a garden and their future is subject to the gardening styles of successive landowners. Little is required to manage these plants and the development of suburbia has not eliminated pollinators, probably because small patches of rocky grassland, unsuitable for development, were left intact at the time the suburbs were planned. The incidence of fire on suburban properties such as this one is a problem where houses have flammable roofing. In addition, cinders and dust have to be cleared from houses for a few weeks after a fire.

Several rare grass aloes are found on strips of rocky grassland in suburbia, for example *Aloe linearifolia* on the outskirts of Margate in KwaZulu-Natal and *A. parviflora* near Botha's Hill north of Durban, also in KwaZulu-Natal. Landowners should be encouraged to retain grass aloes and other naturally occurring plants on their properties as suburbs are established. This would go a long way towards preserving some of the species that have become rare in recent decades. Grass aloes preserved on remnants of grassland in suburbia represent an interesting blend between *in situ* and *ex situ* conservation. Such areas may become significant vehicles for preserving rare grassland flora in the future.

Aloe verecunda has been studied at several localities but most long-term research has been conducted near Carletonville and Westonaria in western Gauteng.

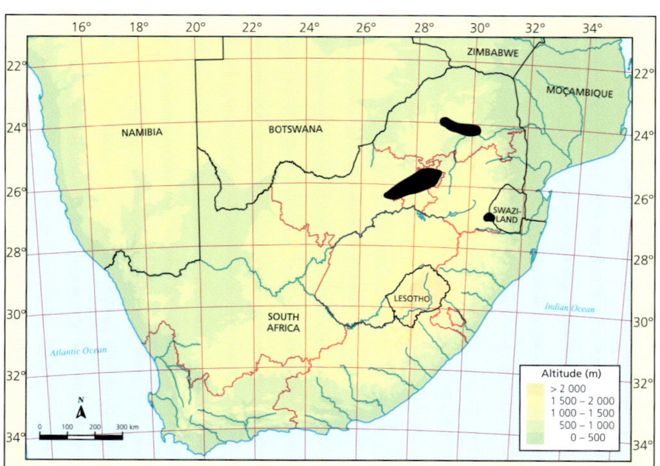

Distribution of *Aloe verecunda*

Southwest of Bothashoop

The Bothashoop border post between Mpumalanga and Swaziland is situated east of Piet Retief. The area has a significant amount of hilly grassland that has escaped afforestation in a region very heavily planted with eucalypts and pines. A long-leaved form of *Aloe verecunda* found here grows amongst rocks near the summit of large south- and southwest-facing hills. These plants are close to *A. vossii* in many respects, particularly their long leaves and the shape of the inflorescences. This rocky habitat is similar to most places where *A. verecunda* grows in western Gauteng and to the little that remains of its habitat around Johannesburg and Roodepoort.

The summit of the hills contain large sheets of exposed rock and are exposed to direct sunlight for the greater part of the day. This area is densely utilised by *Aloe suprafoliata*. However, *A. verecunda* is not found in these hot, arid situations but on the cooler, moist hill slopes.

Aloe verecunda grows mostly in large clumps consisting of many individual stems wedged in between rocks. These situations also harbour a large number of short grass tufts and some bracken (*Pteridium aquilinum*). The habitat remains moist for much of the summer. Rainfall is augmented by frequent mists and aloes growing near seepage areas are kept moist for long periods in summer. This extensive rocky habitat is rare in the Piet Retief area and is not

Right: ***Aloe verecunda*** typically grows on rocky ridges, surrounded by short tufts of grass and the extensive root systems penetrating narrow fissures in the rocks. In these situations, like here in the vicinity of Bronkhorstspruit in Mpumalanga, the plants escape the hottest flames during a grass fire and only their dead leaves are burnt together with the surrounding dry grass.

usually utilised by the other grass aloe species that occur in the area, namely *A. ecklonis, A. hlangapies, A. integra* and *A. kniphofioides*. The hills are too rocky for afforestation with plantations, although alien timber has been planted a mere 20–30 m away from one of the *A. verecunda* colonies. The different populations consist of mature and young plants. The veld is regularly burnt in the winter months since it is situated just outside the perimeter of a fire prevention zone.

The seed set, in common with many other *Aloe verecunda* populations elsewhere, is very erratic from one year to the next. The rocky grassland is well preserved but some of the habitat is being invaded by black wattles (*Acacia mearnsii*), which, if unchecked, can be expected to transform much of this hilly grassland. The proximity of afforestation with its attendant weeds such as bugweed (*Solanum mauritianum*) is also resulting in additional pressures on the *A. verecunda* habitat.

Aloe verecunda is, fortunately, free from invasive alien vegetation in most of its distribution range.

Near Middelburg in Mpumalanga

Aloe verecunda is frequent on a range of hills on the northern outskirts of Middelburg, a town in Mpumalanga. The aloes here grow on south- and southwest-facing hill slopes amongst tufts of short grass and boulders. Some of the plants form multistemmed clumps, particularly where they have taken root in deep pockets of soil between boulders. The grassland is regularly burnt during the rainless winter months and there are a number of young plants in the population. The frequency of winter veld fires has restricted the development of most seedlings to rocky places where they escape the worst effects of the flames.

An extensive rocky ridge runs immediately west of Middelburg in a roughly north-south direction. The eastern and western slopes are comparatively dry and do not appear to house any populations of *Aloe verecunda*. The aloes are often absent from the northern slopes of hills exposed to day-long direct sunlight. Exceptions to this general rule are escarpment habitats in the mist belt, where the aloes occur on the uppermost regions of the northern slopes.

In western Gauteng

The largest numbers of *Aloe verecunda* are still to be found in western Gauteng, the most densely populated province in South Africa. Much of it will be developed and urbanised further in the coming decades. Before the discovery of gold on the Witwatersrand, *A. verecunda* used to occur in a vast semicircle of separated as well as continuous populations from southwest of Carletonville eastwards to Johannesburg and northwards towards Roodepoort and Krugersdorp. About a third of these plants now survive in relatively pristine habitat situated roughly between Carletonville and Westonaria. The preservation of these plants is directly related to deep-level gold mining which has kept out other forms of land use such as agriculture and stock farming. Urban developments on mine property have furthermore been limited to designated areas. The highest concentration of plants occur in the hills east of Carletonville. The aloes grow mostly well away from slimes dams and rock dumps, the by-products of gold mining. Suburban development has taken place on the ridges near Westonaria, but mostly on the northern slopes not usually frequented by *A. verecunda*.

A large population of the aloes has been studied near Westonaria for some 15 years. The plants grow on the summit, southern and southwestern slopes of a large hill with an altitude of almost 1 800 m at its highest point. Most aloes are large and multistemmed, growing in deep soil between rocks or in fissures near the edges of rock sheets. Much of the habitat is situated on the sides of seepage areas that remain moist for long periods after rain.

The plants flower and seed well in summers with good rainfall between November and late January, after the rocky grassland has been burnt the previous winter. Seeding is rather poor in years when few plants in the population flower, but the reverse occurs when there is mass flowering. Large numbers of bright reddish orange flowers serve the purpose of attracting the range of insects and sunbirds (species of *Nectarinia*) that pollinate these flowers.

The deciduous leaves die right back to the stems in the winter but unseasonal rainy weather at any time during the winter months brings the aloes into leaf. This leaf growth is temporary, however, and is soon cut back by severe frosts once the cloudy weather has lifted.

The sharp division in vegetation between northern and southern slopes is clearly marked at this locality. The northern slopes rise steeply and are exposed to strong direct sunlight for most of the day. *Aloe verecunda* is absent from these hot slopes but common only about 20 m from the summit of the hill where it begins to slope to the west and southwest.

Growth cycle and the role of fire

Aloe verecunda commences growth in November during years when the rocky grassland in which it is found has not been burnt. Leaves are produced much earlier, from about mid-September in years when the veld has been burnt the previous winter.

Flowering occurs from mid-November to late February. In years when there is abundant early summer rain, the peak flowering is usually the second half of November and the first half of December. In years when good rains are late, starting in December, the peak flowering season is usually late December and January. Occasional plants may be found in flower until early March. Most of the aloes are in seed from mid-January until early March. The timing of the seeding season depends entirely on the arrival of the first successive good summer rains, which bring the plants into flower.

Seeds are dispersed mainly by wind. Breezes and gusts of wind are frequent throughout the summer on the exposed, rocky hillsides where *Aloe verecunda* usually grows. Most seeds will obtain the chance to germinate only at the beginning of the following rainy season, in November. This applies particularly to years when the seed set is late, in February and March. In this respect *A. verecunda* is no different from the other grass aloes flowering in late summer and early autumn.

Right: ***Aloe verecunda*** in dormancy in the middle of winter. Rain had fallen a few days before and leaf growth had been initiated, soon to be cut back by heavy ground frosts prevalent at this time of year. Here on the upper slopes of the hills near Westonaria in Gauteng the aloes frequently share their habitat with *Adromischus umbraticola*.

G Condy
2008

Fire plays an integral role in the autecology of *Aloe verecunda*. The flames clear the grassland of tall grasses and burn out accumulations of dead aloe leaves deleterious to the health of the plants. Flowering performance is always improved after a winter grass fire. Masses of brightly coloured aloe flowers in an otherwise drab landscape of hilly grassland attract large numbers of pollinators. A good seed set usually follows and the seeds are shed into the environment at a time when there is the thinnest grass cover. This is conducive to good germination should there be average or above average rainfall the following summer. Fires are far more frequent in much of the *A. verecunda* habitat than they were in former times. It is for this reason that most young plants are found amongst rocks where seedlings are sheltered from the hottest burns. In habitats that are burnt only once every two to three or more years, young plants of *A. verecunda* are able to survive in more open situations with deep soil.

Status of wild populations

Aloe verecunda is still quite a common, widespread species. This is attributable to the fact that deep-level gold mining and urbanisation have not adversely affected the role of fire in rocky grasslands nor have pollinators become scarce. The importance of rocky ridges in Gauteng has been formally recognised, implying that further development in these areas will be curtailed. Illicit dumping of building rubble and garden refuse as well as the chronic spread of weeds and invasive alien vegetation continues, however, to degrade habitats. In this respect preservation of grassland within suburbia could play a significant role in conserving this species in Gauteng.

The species does not occur by and large in regions used for producing alien timber and this augurs well for its future preservation. The hills east of Piet Retief are likely to become degraded and eventually transformed by aliens if the habitat in this region is not actively managed.

Cultivation, propagation and horticultural potential

Aloe verecunda makes a good subject for sunny rockeries with deep pockets of soil, in summer-rainfall areas. The plants can also be grown in well-drained soil in containers. A good cultivation mix consists of one part sieved sandy soil and one part silt to which some very well-rotted compost has been added and well mixed in.

Plants should be well watered at intervals of about seven to ten days when they are grown in containers for the duration of the late spring, summer and early autumn months. Aloes grown in rockeries do not have to be watered in seasons with regular and well spaced rains, but should be watered well once every two to three weeks during periods of drought.

This species is easily rooted from cuttings struck in moist, coarse sand. They should be broken off the parent plant and allowed to dry and heal for about two weeks before they are planted them in sand.

Aloe verecunda is easily propagated from seeds sown in deep seed trays or large pots. Seeds can be sown in the same mixture used for cultivating mature plants. They should be lightly and evenly scattered across the surface of the soil and just covered. The mixture in the propagation trays or pots should be kept moist while the seeds are germinating and for the duration of the summer while the young plants are developing. Seeds are best sown in spring, giving the plants the rest of the summer to develop before their winter dormancy.

The plants should be kept dry in winter except for the occasional thorough watering after the winter equinox. This ensures that the plants are in good condition at the beginning of the summer growing season.

Aloe verecunda is a useful subject for gardens where the naturally occurring grassland is preserved. The species can also be effectively and attractively used for mass planting in clumps in gardens with naturally occurring rocky outcrops.

26. *Aloe vossii*

Aloe vossii Reynolds

Aloe vossii is a robust species endemic to the Soutpansberg, the most northerly mountain range in South Africa. The plants occur mostly in the central parts of the Soutpansberg, with some populations along the eastern and western parts of the mountain. The Soutpansberg receives frequent mists in summer and autumn. These mists are heaviest along the uppermost parts of the mountain, decreasing rapidly in frequency and duration on the northern slopes. The aloes occur mostly at altitudes of 1 500–1 700 m, the area receiving the most precipitation in the form of rain and mist.

Flowers of *Aloe vossii* are usually borne in a slightly conical capitate raceme. They are mostly broad at the base and narrow at the mouth. The shape of the inflorescence and flowers are characters that have been used to distinguish these plants from *A. verecunda* (Reynolds 1969: 136–137; Van Wyk & Smith 1996: 294, 296; Glen & Hardy 2000: 16–18). The floral characters are, however, very variable. In this respect, a number of plants of *A. vossii* were grown from seeds collected near Entabeni in the eastern Soutpansberg. Ten specimens of these plants in flower were examined on 8 February 2005. Some of the flowers borne by individual plants were typical of those that characterise *A. vossii*, while others were almost uniformly tubular and indistinguishable from those of *A. verecunda*. A typical inflorescence of *A. vossii* in the sense of the species concept used by Reynolds appears on page 121.

It is clear that *Aloe vossii* is very close to *A. verecunda* in almost all readily visible respects. The only character that is reasonably constant, are the long leaves. However, long leaves are also found on some plants in geographically isolated populations of *A. verecunda* such as those at Athalia east of Piet Retief in Mpumalanga. Some *A. vossii* populations on the Soutpansberg are virtually indistinguishable from *A. verecunda*, except in leaf.

It may be found by future researchers that *Aloe vossii* is an ecotype of *A. verecunda*. For this to be established, a detailed survey would have to be completed along the Soutpansberg. A sample of plants would have to be assessed, drawn from different mountain-summit populations as well as from those that occur less frequently on the upper northern slopes. These plants would have to be compared and contrasted with a representative selection of *A. verecunda* from all parts of its distribution range.

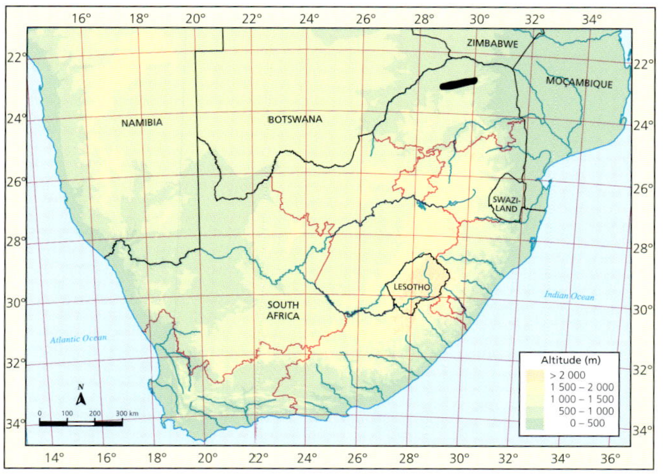

Distribution of *Aloe vossii*

The central and eastern Soutpansberg has been extensively degraded by afforestation, particularly alien eucalypt plantations. Much of the grassland habitat has been destroyed, the remaining parts consisting of small pockets of grassland between plantations. The larger tracts of grassland habitat that remain are often very rocky and lie between extensive areas of indigenous forest and shrubland. The western Soutpansberg is not heavily forested with aliens but as it is drier than the central and eastern parts, it contains limited amounts of habitat suitable for *Aloe vossii*.

The species has been studied at two localities, the summit of the Soutpansberg near Hangklip and the lower northern slopes near the Ingwe Motel. The Hangklip population is at an altitude of about 1 700 m and the Ingwe plants grow at approximately 1 500 m. The two populations are separated by a short distance of some 6 km. The Hangklip plants are typical representatives of the species as described by Reynolds (1936) with long, copiously white-spotted leaves. Many of the Ingwe plants are virtually indistinguishable from *Aloe verecunda* except for specimens with long leaves normally growing in well-irrigated rocky places with the deepest soil.

Near the Hangklip

Near Hangklip *Aloe vossii* is confined to rocky outcrops on the summit of the Soutpansberg. The plants grow amongst grass and bracken (*Pteridium aquilinum*), with their roots tightly wedged into rocky crevices. These sites are not ideal habitat for the aloes since they are surrounded by thickets of indigenous bush and the bracken becomes very dense when it is subject to frequent burning. The bracken and bush limit the chances of *A. vossii* to occupy more open habitat and also the variety of locations that the plants are able to utilise near the Ingwe Motel.

There is furthermore encroachment by alien weeds that have degraded parts of the habitat. Seeds liberated by *Aloe vossii* are afforded little opportunity to germinate and develop into young plants since the surrounding vegetation is too dense. It is likely that, as the years progress, some of the smaller populations of near the Hangklip will disappear under dense bracken and indigenous bush.

The effects of heavy afforestation on the grassy southern slopes of the Soutpansberg have eliminated most of the habitat of *Aloe vossii*. In many cases, the plants on the summit of the mountain have now been confined to the least suitable habitat, namely the transition area between grassland and indigenous forests.

The Soutpansberg is subject to frequent and prolonged droughts, but *Aloe vossii* manages to grow and even flower under such conditions as a result of moisture provided by periodic dense mists.

Near the Ingwe Motel

Aloe vossii grows on a rocky spur of the northern slopes of the Soutpansberg close to the Ingwe Motel. This area lies just within the mist belt and consists of rocky grassland above stretches of indigenous forest and bush. The plants grow amongst rocks and tufts of short grass, in pockets of soil at the edges of sheets of exposed rock and in fissures of rocky outcrops filled with soil. The species is at its most numerous in rocky grassland and least plentiful on rocky outcrops. The aloes form clumps consisting of several stems when they grow in rocky grassland but are only sparsely branched when they occur in very rocky situations with shallow soil.

Plants of all sizes are found, ranging from large clumps to young plants with only a few stems. The species is locally well represented in its rather limited habitat, most of the mountainside containing copses of trees with dense undergrowth. There is a large stretch of open, short grassland close to the rocky areas where *Aloe vossii* occurs but the plants do not seem to grow here. The optimum terrain for this species consists of hilly, rocky grassland epitomised by the habitat occupied by the plants near the Ingwe Motel.

This section of the Soutpansberg is too rocky for afforestation and plants growing here are consequently likely to survive in the long term. The species is, however, relatively scarce on the northern slopes of the Soutpansberg owing to drier conditions than those found on the summit of the mountain range. Plants at this locality receive enough mist during periods of drought to bring them into leaf and flower.

Growth cycle and the role of fire

Aloe vossii is a summer-growing species, usually producing new leaves in November, the beginning of the rainy season in the Soutpansberg. The plants grow actively during late November, December and January. Flower buds start to appear just before the second half of January and the peak of the flowering season is generally the last few days of January and the first half of February. During very dry years most plants produce only a few leaves but they do manage to flower after a few dense mists or light rain.

This species is often semideciduous in winter and may still be in full leaf in June if there has been abundant late rain in the Soutpansberg. The semideciduous habit is retained in cultivation and characterises plants grown in Johannesburg with a much colder winter climate.

Seed is set mostly in March and early April. Aloes growing on the summit of the Soutpansberg are often shrouded in mist for long periods in the late summer and early autumn. The result is that seed capsules may take a long time to split open fully. Seed capsules still retaining some ripe seeds can be found as late as the beginning of June.

Fire periodically clears the northern grassy slopes of the Soutpansberg and is necessary for the health of plants growing in rocky grassland. Most plants of *Aloe vossii* come into flower in late summer and early autumn after a grass fire the previous winter. Fires are actively prevented in most timber-producing parts of the Soutpansberg. However, in other nearby areas, particularly around Bluegumspoort in the west-central part of the range, they are more frequent. In areas that are burnt too frequently bracken (*Pteridium aquilinum*) has replaced some of the grass cover. These dense stands of bracken compete directly with *A. vossii* for moisture and light. Where the cover of bracken is taller than the aloes they prevent the aloes from flowering.

Aloe vossii growing on the Soutpansberg is in the fortunate position that only part of its habitat has been afforested with aliens. Flowers are regularly pollinated from one season to the next as partial afforestation of the mountain slopes with aliens has not reduced the number of pollinators. In this respect the situation differs from that encountered in hilly and mountainous parts of Mpumalanga, heavily afforested with commercial timber plantations.

Status of wild populations

The species is endemic to the Soutpansberg and has declined as a result of habitat degradation. Most of its distribution range on the southern slopes of the central parts of the mountain range has been destroyed by timber plantations. Plants in marginal habitat on the summit of the mountain are subject to habitat alteration by bracken, usually brought about by too frequent veld fires.

Aloe vossii is one of the few South African species of *Aloe* to be listed on CITES Appendix 1. Its status may, however, have to be reviewed if the plants are in due course described as a subspecies or variety of *A. verecunda*. It is also possible that *A. vossii* is only an ecotype of *A. verecunda* representing one of the geographically isolated populations of this widespread and variable species.

Landowners on the summit of the Soutpansberg or its upper northern slopes should be encouraged to preserve populations of *Aloe vossii* on their property. An increasing number of farms on the Soutpansberg have recently become nature reserves. This form of land use, which is often based on ecotourism, is likely to prevent further decline in *A. vossii* populations.

Cultivation, propagation and horticultural potential

Aloe vossii is one of the few grass aloe species that thrive in cultivation. The plants flower regularly year after year provided that they are grown in areas exposed to direct sunlight or very lightly dappled shade. They grow best when planted in sunny garden beds that are not forked over or disturbed. Gently sloping, well-drained situations are ideal, allowing the aloes to develop their extensive root run.

Plants also grow well in large containers but care has to be taken that the roots do not suffer from root mealybug infestations. *Aloe vossii* requires plenty of free air circulation and should not be grown close to a wall or in confined spaces. The species grows quickly and easily from seed, taking about three to four years to reach flowering size. The plants form multistemmed clumps after about five to six years. *A. vossii* sets seed readily in cultivation and the yield is improved with hand pollination. This species hybridises readily with other aloes and care has to be taken to ensure that plants do not grow near other aloe species that flower at the same time, if pure seed is required.

Aloe vossii is not very widely cultivated despite the fact that it is one of the most ornamental grass aloes and also one of the easiest to grow. The two most attractive and successful plants for cultivation purposes are those that occur naturally in the Hangklip area and around Entabeni towards the eastern extremity of the distribution range of this aloe on the Soutpansberg. The Hangklip plants have very long, copiously white-spotted leaves, mostly reaching a length of 35–45 cm, and thick stems up to about 15 cm high. They are ideal for lightly shaded positions where they flower well. The Entabeni plants have shorter, thinner leaves and sucker freely. They grow well in places receiving direct sunlight for much of the day.

Right: ***Aloe vossii*** in habitat on the western slopes of the central Soutpansberg in Limpopo Province. The habitat consists of boulder outcrops and rocky slopes interspersed with dense tufts of short grass. Plants in this colony conform to the descriptions of *Aloe vossii* and *A. verecunda*.

In terms of its flowers, ***Aloe vossii*** growing on the Hangklip in the central Soutpansberg in Limpopo Province is often indistinguishable from *A. fouriei*.

The best time of the year for sowing seeds is early summer or early autumn. Seeds should be lightly and evenly scattered across the surface of deep propagation trays or large, deep pots. The ideal sowing mixture consists of half fine silt or fine gravel, and half sandy soil to which a small amount of well-rotted, finely sieved leaf mould has been added. The seeds should be lightly covered with soil and well watered until they have germinated. Seedlings should be kept moist but not too damp. This species is very prone to damping off and seedlings should be regularly sprayed with a copper-based fungicide during hot or humid weather. Seedlings should be transplanted towards the end of their second growing season.

Aloe vossii is easily grown from cuttings, which should be manually broken off the parent plant and allowed to dry for about seven to ten days. Cuttings should be struck in coarse, gravelly sand.

27. *Aloe woolliana*

Aloe woolliana Pole Evans

Reynolds (1969: 129) expressed the view that *Aloe woolliana* may merely be a robust form of *A. chortolirioides* since the two species share a number of morphological similarities. Glen & Hardy (2000: 12–13) formally recognised this perspective, instating *A. chortolirioides* var. *woolliana* in their revision of the genus *Aloe*.

Aloe woolliana is, however, retained as a separate species in this work since there are enough differences between the two plants to keep them separate—aspects of the autecology of the two plants, their morphology and their current status in South Africa.

Aloe chortolirioides produces annually on average 10 to 20 leaves only 1–2 mm wide, whereas *A. woolliana* mostly has five to seven annual leaves which are generally 6–8 mm wide. *A. chortolirioides* frequently rebranches whereas *A. woolliana* suckers from the base, producing large clumps of stems. *A. woolliana* favours east- and southeast-facing upper grassy cliffs where this habitat is found in preference to rocky, hilly grassland frequented by *A. chortolirioides*. The equivalent cliff-face habitat utilised by *A. woolliana* occurs in the mountains near Barberton, the home of *A. chortolirioides*. It is, however, not a habitat favoured by *A. chortolirioides*.

No populations where the two species intergrade were found during the 20 years of fieldwork on which this book has been based. Should such populations be located by future workers in the field, such data would provide one reason to prompt re-assessment of the separation of these two plants. The autecology of *Aloe chortolirioides* is integrally linked to fire cycles, whereas *A. woolliana* is not as dependent on fire, particularly with respect to the abundance of flowers and timing of the flowering season.

Aloe woolliana has been severely affected by afforestation in the Lydenburg and Sabie Districts as well as in the area around The Brook, all parts of Mpumalanga. Only one significant population of *A. woolliana* was located near The Brook and this interesting ecotype has all but disappeared from the southwestern part of its distribution range.

Aloe woolliana is one of several species of which the habitat is shrouded in heavy mist for days on end in the summer months. The rocky cliff faces frequented by the plants are very wet during rainy and misty weather but dry out when it clears. The long, thick, succulent roots that characterise the species are perfectly adapted for drawing up moisture and anchoring the plants in grassy swards on cliff faces.

Four populations of this species have been studied, one close to Sabie, another in the mountains near Lydenburg and two widely separated populations, one in the Carolina District and the other on the Steenkampsberg. The two last-mentioned localities represent the extreme western and southwestern parts of the distribution range of this species.

Near The Brook, southwestern Mpumalanga

A most unusual ecotype of *Aloe woolliana* is found near The Brook in Mpumalanga. The plants grow in shallow soil pockets on huge sheets of exposed rock, either in the open in strong direct sunlight or amongst tumbled boulders on

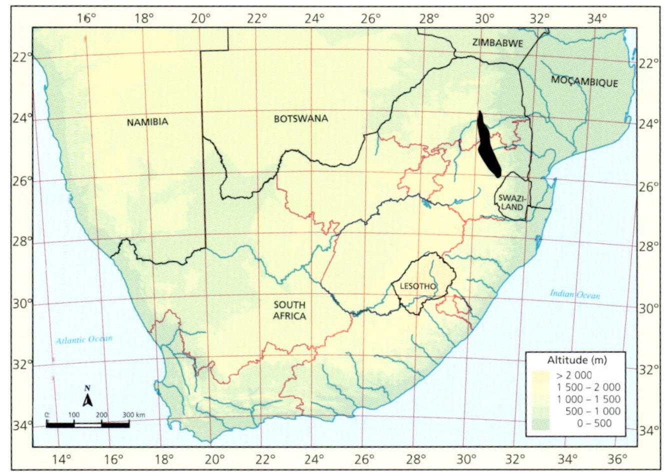

Distribution of *Aloe woolliana*

the rock sheets. This habitat is exposed to periodically very wet and very dry conditions. At times the aloes stand in shallow pools of water for a few days in summer. The leaves die back in winter, forming mats shielding the plants from the desiccating rays of the sun. The individual stems are only 6–8 cm high. The plants often form extensive clumps of 30 stems or more, most of them suckering from the base during prolonged wet weather in summer. This habitat is shared with a dwarf ecotype of *A. minima* that grows in shallow soil amongst short grass on the rock sheets and also in surrounding rocky grassland.

This locality is one of the few places where *Aloe woolliana* occurs commonly on north- and northwest-facing slopes, probably because the habitat is at a high altitude and very moist during much of the summer.

On the Steenkampsberg, western Mpumalanga

Where *Aloe woolliana* occurs on the summit areas of the Steenkampsberg, the plants usually grow on southwest- or east-facing ledges of small cliffs, which are frequent on the mountain summit. The plants are rather sparsely and widely scattered. Several multistemmed clumps sometimes occur in close proximity to one another, frequently sharing their habitat with *Eucomis vandermerwei*. The mountaintop habitat is subject to low temperatures, frost and even occasional snow in winter. The summers are cool, with frequent mists. *A. woolliana* usually comes into flower in early October on the Steenkampsberg, at a time when the plants on the mountains near Lydenburg and Sabie are starting to form seeds.

The distribution of *Aloe woolliana* on the Steenkampsberg is natural, representing the sum total of habitats in the rocky mountaintop locality occupied by the species. Afforestation has not taken place on the Steenkampsberg and this habitat affords good opportunities for the study of *A. woolliana* under relatively pristine conditions.

On the Drakensberg, southeast of Lydenburg in Mpumalanga

Aloe woolliana is still well represented along the Drakensberg Escarpment near Lydenburg where the plants occupy the full range of habitats frequented by this species. They are found amongst boulders in rocky grassland above cliff faces, on grassy cliff ledges and at the edge of seepage areas on cliff faces. They are most common at the edges of seepage areas on the cliffs where they usually form large multistemmed clumps that protrude from the cliff faces. These sites are very well irrigated and several individual aloes are sometimes clumped together, creating the impression that they form a single plant.

This area is heavily afforested with pine trees, above and below the rocky cliff face habitat. It has brought about changes to the natural fire regime and also appears to have decreased the abundance of natural pollinators. A stretch of some 3 km of cliff face habitat was examined on 21 September 2004, the peak of the flowering season of *Aloe woolliana*. No pollinators were seen on the flowers during the late morning and early afternoon, when they are usually the most active. The pollinators, mostly honeybees (*Apis mellifera*), were attracted in large numbers to clumps of Natal bottlebrush (*Greyia sutherlandii*) in full flower but avoided the widespread, more sparse *A. woolliana* flowers.

Very few young plants were found, typical of areas heavily afforested with alien timber trees. Periodic fires stimulate mass flowering which in turn renders the plants more visible and attractive to pollinators. In addition, flowers in strong sunlight are often the most attractive to pollinators and most plants of *Aloe woolliana* grow on east- and south-east-facing cliffs where they receive largely early morning sunlight. These situations are often in shade from about noon during the peak of the flowering season.

Near Sabie in Mpumalanga

Near Sabie, *Aloe woolliana* inhabits small pockets of rocky grassland. Most of the existing populations are not viable in the long term since they are completely surrounded by commercial timber plantations and widely separated from one another. One colony is never burnt as it occurs in the middle of a pine plantation. This leads to year after year of substandard flowering. Winged insect pollinators avoid these aloes surrounded by pine forests and there is consequently little, if any, regeneration from seed.

Growth cycle and the role of fire

Aloe woolliana is a spring- and summer-growing species. The plants are dormant from late May until late August around Lydenburg and Sabie and from late April until late September on the Steenkampsberg and near The Brook. These two last-mentioned areas are much colder and the beginning of dormancy is therefore earlier and the commencement of the growing season later.

Grass fires stimulate mass flowering, which is often a month or more earlier than the usual September and October flowering peak in years when there are no fires. The typical flowering season in years without fire is reflected by the data collected on a field trip to the Lydenburg and Steenkampsberg colonies on 21 September 2004. The plants on the Drakensberg escarpment were mostly at the peak of the flowering season, with some late plants in heavily shaded positions still in bud. On the Steenkampsberg, bud formation had just been initiated on some of the stems.

Aloe woolliana flowers regularly without the prior burning of its habitat, but the flowering performance of individual plants is much improved after a thorough burn. The frequency of fires has been much reduced over the last 40 to 50 years owing to the establishment of the timber industry in most of its habitat. Fire prevention measures are implemented in most parts of Mpumalanga where the aloes occur. There is, as a consequence, a tendency for plants to set seed in the most open habitat, uncluttered by moribund grasses—mainly rocky places on cliffs and open patches amongst boulders on the summits of mountains.

Status of wild populations

Much of the optimum habitat for *Aloe woolliana* in Mpumalanga is under commercial tree plantations. The most viable populations of this species are along the Drakensberg Escarpment near Lydenburg. There are some preliminary indications that the lack of enough periodic fires and changes to the pollination biology of the plants may have some effects on the population dynamics of the species in the long term. Plants near Sabie provide good examples of how extensive timber plantations destroy the long-term viability of *A. woolliana* populations in such areas.

The colonies on the Steenkampsberg occur in the least disturbed habitat and probably rank amongst the best for long-term preservation initiatives concerning this species. They will also require the least intervention for long-term management of this grass aloe. The interesting ecotype near The Brook still occurs in one large, healthy population in open grassland that has escaped afforestation.

Cultivation, propagation and horticultural potential

Aloe woolliana is easily cultivated in a well-drained, gritty medium consisting of a third coarse, sieved sandy soil, a third river sand and a third very well-rotted, sieved leaf mould or well-rotted, sieved veld grass compost. Plants thrive in large containers in open, sunny positions with much free air circulation. They should be well watered during the period September to March, which is the main growing season, and only lightly and occasionally during winter.

The aloes flower year after year without burning, but flowering is usually rather sparse. Flowering performance can be increased by placing dry grass harvested from the veld around the aloe stems towards the end of winter and then igniting it. The aloes should be well watered about once a week afterwards until they come into flower. Another equally successful method is to grow clumps of veld grass together with the aloes and to burn them at the end of each winter. Care has to be taken, however, that the grass does not usurp all the pot space and it may be necessary to remove tufts every few years when the grass has seeded itself.

Aloe woolliana is particularly attractive when planted amongst ornamental rocks in tall terracotta containers with some veld grass. This form of display is an imitation of the natural conditions in the rocky habitat where the plants are found.

Propagation is best achieved by means of stem cuttings, which should be rooted in sand in spring or autumn. The sand should be kept moist throughout the rooting process.

Right: ***Aloe woolliana*** growing on an escarpment cliff near Lydenburg in Mpumalanga Province. The aloes are abundant in places and their cliff-top habitat is wedged in between enormous, virtually uninterrupted plantations of eucalypts and pines. In former times the plants occurred on cliffs and in the surrounding rocky grassland.

The escarpment cliffs along the Drakensberg near Lydenburg are home to healthy populations of *Aloe woolliana*. The plants typically grow at the edges of seepage areas in grassy swards.

Roots have usually developed well after about two months and the plants can then be replanted in the containers where they will be permanently grown.

Seed is rather erratically and sparsely produced in cultivation but seed set is much improved by hand pollination, particularly if pollen is gathered from several plants. Seeds should be lightly and evenly scattered across the same mixture used for cultivating adult plants. They should be just covered with soil and kept moist while germinating and for the next few months, until the seedlings have developed small trunks. The young plants are best transferred to the containers in which they are to be grown as mature plants towards the end of their second season.

Aloe woolliana on the Steenkampsberg has striking bright pink flowers and is very floriferous. These plants could become popular in horticulture and consequently attempts should be made to see if they can be mass-produced by tissue culture.

28. *Chortolirion angolense*

Chortolirion angolense (Baker) A.Berger

Chortolirion angolense shares most of its growth habit, habitat preferences, stimuli for flowering, and seed dispersal patterns with grass aloes. The structure of its underground bulb is very close to that of the three bulbous South African grass aloe species *Aloe inconspicua*, *A. kniphofioides* and *A. modesta*.

Chortolirions, in common with grass aloes in South Africa, form a component of the deciduous flora in a grassland ecosystem regulated by fire cycles.

The variables concerning the placement of *Chortolirion* in a hierarchy of genera are the subject of some debate. This is to be expected since the observable variation in *Chortolirion*, and some of the biological processes central to its life cycle, are not comfortably accommodated within the binomial system. These matters are described and examined in this chapter. In addition, the same detailed information about the plants in their natural habitats that has been discussed under the species accounts of the different grass aloes, is provided for *C. angolense*.

A recent paper by David Cumming, based on intergeneric hybridisation experiments lends support to the proposition that *Chortolirion* is close to *Haworthia* subgenus *Hexangulares* (Cumming 2005). Intergeneric hybridisation experiments were conducted between *Chortolirion* and some grass aloe species as well as between *Chortolirion* and some *Haworthia* species. The conclusions reached in David Cumming's paper are at variance with those presented by Treutlein *et al.* (2003) who demonstrated a relatedness between *C. angolense*, *Aloe boylei* and *A. verecunda* with respect to their DNA sequencing.

These kinds of viewpoints all contribute to a re-defined concept of *Chortolirion* and its future placement in a hierarchy of genera as the techniques for the production of botanical knowledge change and evolve. It has to be remembered that the genus *Chortolirion* is a concept defined by the function it performs in the organisation of botanical knowledge. Contributions to the debate of what comprise the boundaries for circumscription of the genus will, accordingly, be presented at various levels of abstraction which may or may not share common ground.

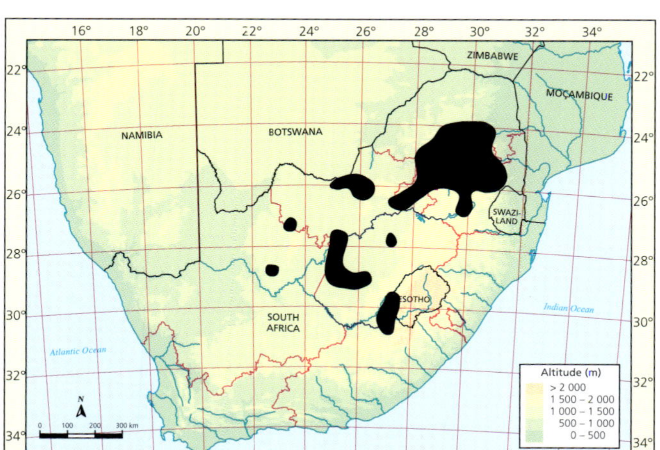

Distribution of *Chortolirion angolense*

The plants that fall within the genus *Chortolirion* are also a problem. The binomial system is unsuited to dealing with certain defining characteristics that separate the plants from one another. This interesting set of circumstances is discussed in detail in Craib *et al.* (2004: 1–6). The details of this paper are summarised here, with additional comments elicited from the conclusions reached by the authors.

The concept of *Chortolirion* has included a genus with species, the transfer of species in the genus to the genus *Haworthia* and the reduction of species in the genus *Chortolirion* to one species. *Chortolirion* is currently a monotypic genus. If *Chortolirion* is retained as a genus by future workers, two difficulties present themselves with respect to the organisation of knowledge—variation within *Chortolirion* and the time of the year at which the plants flower. *Chortolirion* populations have distinctive spring or autumn flowering seasons.

In the past there was a tendency for *Chortolirion* species to be described on the basis of extreme variants with respect to their morphology. This kind of situation is nothing new, particularly where populations of various plant species have become geographically fragmented by afforestation and habitat destruction. The difference in the case of *Chortolirion* is that species were published before detailed research about their distribution ranges had been done. There is some recent evidence to suggest that certain variants in *Chortolirion* are relatively uniform across a wide geographical area. In this respect, *C. angolense* found near Belfast in Mpumalanga shares morphological characters with plants found to the west and southwest near Bronkhorstspruit and Carletonville in Gauteng and Delmas, also in Mpumalanga.

Chortolirion angolense growing near Odendaalsrus and Welkom in the Free State has the extraordinary character of producing a rosette of broad, spirally twisted leaves like certain populations of *Gethyllis villosa*. These plants have so far been found in only three populations and are likely to be scarce owing to their specific habitat requirements, which include deep, soft, sandy soil.

Entities or forms with these characters, collected during the research for this book, suggest that a concept of *Chortolirion* with subspecies or varieties may merit further attention. If *Chortolirion* is reinstated with species, subspecies or varieties there will inevitably be the problem of boundaries between different entities.

Recent research (Craib *et al.* 2004) has indicated that *Chortolirion* has two distinct flowering times, namely spring and early summer and autumn. The data collected indicated that spring-flowering chortolirions do not flower in autumn and vice versa. Flowering times can be used to formulate a model around which data concerning variety in *Chortolirion* can be organised. The problem in this instance is that the binomial of the species cannot comfortably accommodate the two phenological variants. This situation should certainly facilitate a reassessment of the infraspecific taxonomy of the species.

There are some indications that *Chortolirion* populations with unique physical attributes or uniformity across geographical space, can be grouped under the characters of spring or autumn seasons. The possibility exists, certainly according to logical reasoning, if nothing else, that phenology could be an overarching model within which species or infraspecific taxa could be reinstated.

This résumé of recent developments concerning the development of knowledge about *Chortolirion* indicates that the subject is far from clarified or settled. The view of Craib *et al.* (2004), namely that *Chortolirion* is a monotypic variable genus with distinct flowering times, is upheld here and is used for the discussion that follows. It is anticipated that revised concepts of *Chortolirion* will be presented in the future as a result of innovations in the techniques used for constructing botanical knowledge.

Spring-flowering *Chortolirion angolense*

Spring-flowering plants of *Chortolirion angolense* flower mainly between late August and early November, with a peak in September and October. They usually grow in short grassland, mostly flat or gently undulating. Many of the largest populations used to occur around Johannesburg and Pretoria in Gauteng. They have been progressively exterminated by urban and suburban expansion, illicit dumping of waste material and habitat degradation brought about mostly by annual and perennial alien weeds. The conditions in the grassland surrounding Pretoria and Johannesburg have, for a long time, permitted the expansion of *Chortolirion* populations. Winter grass fires are frequent and many areas of open veld have only been moderately grazed by a few horses and cows where they fall within areas where smallholdings have been developed. Large tracts of grassland have not been grazed by domestic stock, particularly on the outskirts of suburbia. In rural grasslands, *C. angolense* is subject to different kinds of conditions and processes from those occurring in the habitat around Johannesburg and Pretoria.

In the Hillsview and Honingklip areas of the northwestern Witwatersrand in Gauteng

The Hillsview and Honingklip areas lie east of Krugersdorp in Gauteng. In recent years there has been extensive suburban development but prior to this the region consisted mostly of smallholdings and deciduous fruit farms. Some of the largest colonies of *Chortolirion angolense* in Gauteng used to occur in these areas, in short grassland at the base of a range of hills. The plants were most plentiful in short open grassland at the bottom of the hill slopes. They also occurred on the hillsides, but were much less plentiful. Grazing was virtually absent and veld fires were a regular occurrence in the winter months. The grassland was truncated by smallholdings and farms but this made little difference to the biological processes required to sustain the plants. Flowers were well pollinated year after year and good seedling recruitment was evident in years of average or above-average rainfall.

In 2003, large townhouse developments destroyed the greater part of the habitat. The plants have now been confined to the rocky hill slopes in areas where developments have not taken place. This is the least suitable habitat for *Chortolirion angolense* but some plants have remained in the area where they used to be more generally distributed.

Near Odendaalsrus and Welkom in the Free State

Odendaalsrus and Welkom are two towns situated about 11 km apart on the Free State gold fields. They are surrounded by short grassland rich in bulbs and caudiciforms. Some of the grassland close to the towns is still well preserved but in most areas the veld has been degraded to varying degrees. Huge parts of natural habitat have been destroyed by mine dumps and maize fields. Many patches of remaining grassland are not grazed by domestic stock and the veld is burnt regularly each winter.

Plants of *Chortolirion angolense* near Odendaalsrus and Welkom have a number of significant characters that separate them from other populations of *Chortolirion*. The plants have broad, erect or prostrate, spirally twisted leaves, often reaching a width of 5 mm or more. The chortolirions have exceptionally long roots, usually reaching a length of 10 cm or more. They usually grow in clumps of five to 40 tightly packed bulbs when they occur in veld that has been little degraded or disturbed. This habit of forming large clumps is likely to be unique since in many other *Chortolirion* populations studied over some 20 years bulbs were only found to occur singly.

The plants appear to be restricted to areas of deep, soft, sandy soil and have not been found in places where there are heavy loams, which are frequent in this part of the Free State. The chortolirions are currently known only from three populations, one near Welkom and two near Odendaalsrus. The plants may also occur around the town of Wesselsbron and also further northeast towards Bothaville. They have not been found to the northwest in the Makwassie, Leeudoringstad and Wolmaransstad areas, despite extensive searches in similar habitat.

Chortolirion angolense near Odendaalsrus and Welkom flowers year after year in late November and the first half of December. In years of drought, fewer plants flower and the timing of flowering is later, peaking in mid-December and even extending into January. The data collected at two colonies, one near Welkom and another north of Odendaalsrus, indicate how quickly the plants decline when degradation of the veld begins.

Colony north of Odendaalsrus

North of Odendaalsrus, *Chortolirion angolense* is found in gently undulating, short grassland about 1 km^2 in extent. The grassland is wedged in between a mine dump and maize fields and also contains a disused stone quarry.

The chortolirions grow in two small sections of the veld, both with deep sandy soil. They have not been found in sandy loam, which is the most common soil type in this area. The largest group of plants is scattered over an area of some 300 m^2. The chortolirions are found mainly in groups of three to 50 tightly packed bulbs. They also occur as individual plants. The largest numbers of bulbs are found on flat ground or in slight depressions, which remain the wettest for the longest periods after rain. These situations afford the best opportunities for germination of seeds. Many

Right: Spring-flowering ***Chortolirion angolense*** in its rocky grassland habitat near Irene in Gauteng. The veld had been burnt in the dry winter, stimulating the plants to come into flower in the early spring. Chortolirions often share their habitat with *Euphorbia clavaroides* var. *truncata*.

The habitat of autumn-flowering *Chortolirion angolense* near Bronkhorstspruit along the boundary of Gauteng and Mpumalanga. The plants grow in short grassland, frequently near rock outcrops and sheets of exposed rock. This habitat is being invaded by eucalypts and black wattles.

plants grow in direct sunlight in open, sandy patches and not amongst grasses. This habitat is very rarely recorded for *Chortolirion*.

The second smaller group of chortolirions occurs close to the disused quarry, some of them growing amongst rocks left behind after quarrying operations. The plants share their habitat with *Brachystelma stenophyllum*, a caudiciform also found in parts of the veld near Odendaalsrus with deep sandy soil.

The illustration for this book was drawn at the largest colony on 16 December 2004, a dry year with erratic rainfall. The plants had just started to come into flower after the first good summer rains had fallen a few weeks before.

The tips of the *Chortolirion* leaves in both colonies had been eaten by locusts and beetles. The same was true of the emerging flower stems. Some of the bulbs were sprouting more stems after the first ones had been nibbled down by insects. Much of the surrounding grassland has been destroyed by maize fields and little available habitat remains.

Colonies west of Welkom

The chortolirions used to be common in deep, sandy soil in a patch of open grassland west of Welkom. This locality is another small remaining segment of the grassland habitat hemmed in between maize fields, an airport and a recreation facility. The plants used to be plentiful in open, sandy patches and were usually encountered in clumps of tightly packed bulbs. Individual plants were also frequent particularly in drier parts of the habitat. Most clumps of bulbs occurred along drainage lines or at the edges of a shallow pan that is temporarily inundated during rainy weather.

The habitat was, and still is, rich in bulbs and caudiciforms. *Brachystelma dimorphum* subsp. *gratum* is found sparsely while *Pterodiscus speciosus* is very common. Various bulbs are frequent, particularly *Ammocharis coranica*, various species of *Dipcadi* and also two species of *Urginea*. The habitat has been gradually degraded over the last ten years. Sections of the veld have been silted over and a long trench had been dug in one section of the habitat. The soil from the trench has been dispersed after heavy rainfall, adding to the siltation problem. Sand has been removed for building purposes, leading to arid patches of compacted, uneven soil and more recently illicit dumping of building rubble has destroyed several parts of the habitat where the chortolirions used to occur.

A survey of the habitat on 13 December 2004 indicated that few chortolirions remained and that there was virtually no recruitment of seedling plants. This situation contrasts markedly with the colonies north of Odendaalsrus where numbers of young plants are present in parts of the veld that have not been degraded or disturbed.

Near the Klerkskraal Dam

The Klerkskraal Dam is situated on the upper reaches of the Mooi River just within the North-West Province close to the border with Gauteng. The surrounding area consists of gently undulating, short grassland. Most of the flatter grassland areas have been converted to maize fields and the remaining parts are used for cattle ranching. The chortolirions are found mainly on low stony outcrops where they grow amongst rocks and short grass. They share their habitat with other bulbs and caudiciforms adapted to conditions where there is shallow, rocky soil, such as *Brachystelma nanum*. The plants may have been more widespread in the open grassland before the land was used for cattle ranching as they are susceptible to trampling by large animals. Small chortolirions in particular are readily destroyed and most seedlings are therefore found on the rocky ridges.

Chortolirion angolense flowers very early in the season near the Klerkskraal Dam when the veld has been burnt during the previous winter, usually in late August and early September. In years when there have been no winter fires, flowering is later and more sparse, usually late in September.

Growth cycle and the role of fire

Spring-flowering forms of *Chortolirion angolense* respond directly to burning of their grassland habitat during the winter months. Winter grass fires stimulate mass flowering in the late winter and early spring. Many populations of *C. angolense* flower before the emergence of new leaf growth, after the veld has been burnt. The chortolirions flower later,

Right: ***Chortolirion angolense*** found around Odendaalsrus and Welkom in the Free State is characterised by broad, spirally twisted leaves and is often clump-forming. These plants were growing in a segment of grassland near Odendaalsrus, wedged in between maize fields and mine dumps.

J Condy

The rocky grassland habitat frequented by the autumn-flowering form of *Chortolirion angolense*, near Middelburg in Mpumalanga, in late July 2005 after a winter grass fire. The chortolirions grow widely scattered in short grass amongst the rocks along the full length of the ridge.

usually in October, where the veld has not been burnt. Fewer plants tend to come into flower and the flowering performance tapers off when there have been no fires in a particular habitat for several years in succession.

Regular fires, often on an annual basis around Johannesburg and Pretoria, and an absence of grazing animals ensure the recruitment of good numbers of seedlings to the different populations. Young bulbs are not affected by regular burning, but they do succumb to trampling by domestic stock.

Urban developments and stock farming appear to have had little effect on the pollination biology of spring-flowering *Chortolirion angolense*. Good seed sets appear year after year. Seeds are dispersed in October and November, mostly by water runoff after rainstorms and also by wind. Groups of chortolirions containing plants of varying ages are often found, the result of most seeds settling and germinating around the parent plants.

Autumn-flowering *Chortolirion angolense*

Autumn-flowering chortolirions start to come into flower about six to eight weeks after the southern summer equinox. This is a time of gradually decreasing day length and high moisture levels in the soil. The peak of the flowering season is usually from mid-February to early March, with occasional plants flowering at the end of January or late in March.

The autumn-flowering plants, like their spring-flowering counterparts, are most abundant in areas where there is no regular and heavy grazing. Where the veld is heavily grazed, autumn-flowering plants of *Chortolirion angolense* are mainly inhabitants of rocky grassland. So far, spring- and autumn-flowering plants have not been located in the same grassland. The closest distance between the two forms currently recorded is about 20 km.

Plants on the outskirts of Middelburg, Mpumalanga

Chortolirion angolense is frequent on rocky ridges flanking the town of Middelburg. This habitat is wedged between the town and a large, expanding informal settlement. The plants grow in rocky grassland and many of them are particularly large and robust. The flowering stems may reach a height of 2 m but those of about 1 m are more frequent.

These chortolirions come into flower in the second half of February. Flowering performance is at its best after a grass fire the previous winter followed by good rainfall in the summer months preceding the flowering season. The plants rarely come into flower during a drought.

This is one of the few localities where *Chortolirion angolense* is well represented on rocky ridges. The main reason is that there are deep sandy patches amongst the rocks conducive not only to the establishment of the large bulbs of these plants but also the germination of seeds and development of seedlings. The pockets of deep sandy soil remain moist for long periods after rain, particularly if they are shaded by rocks and grass. The habitat is being degraded by annual and perennial weeds and the illicit dumping of building rubble and garden refuse.

Searches in similar areas around Middelburg have failed to yield further records of these chortolirions, amongst the most robust ever recorded.

Right: Autumn-flowering ***Chortolirion angolense*** in the vicinity of Bronkhorstspruit near the border between Gauteng and Mpumalanga is characterised by thin, wiry leaves and pinkish buff-coloured flowers.

Colonies northeast of Bronkhorstspruit near the border between Gauteng and Mpumalanga

The largest colonies of autumn-flowering *Chortolirion angolense* are located in hilly grassland north of Bronkhorstspruit. The plants here number several hundred mature specimens. They grow in flat grassland with short cover, on gently sloping grassy slopes with large boulders and in shallow sandy soil at the edges of large sheets of exposed rock. The plants grow close to enormous concentrations of the recently described *Frithia minor*, a leaf succulent partial to shallow sand over sheets of exposed rock.

The chortolirions flower best after grass fires have burnt the veld the previous winter. Their flowering performance decreases when there are no veld fires, particularly in parts of the habitat where the plants grow amongst long grass. Plants that grow in sandy soil with little grass cover, at the edges of sheets of exposed rock, flower regularly from one year to the next, irrespective of grass fires.

During the period the plants have been studied, about eight years in all, the veld has hardly been used for grazing livestock. As a result, there has been good recruitment of young plants, particularly in years with rainy summers. The chortolirions flower from early February until mid-March, with occasional plants flowering as early as the last week in January.

The grassland where the chortolirions occur is well preserved. Much of this grassland could, however, easily be destroyed by extending the nearby maize fields. Part of the habitat has been used for periodic mining of quarry sand, with its associated dumping and disturbance of topsoil. This has led to a proliferation of annual and perennial weeds which have degraded some of the habitat. A more insidious problem is the extensive spread of black wattles (*Acacia mearnsii*) into a large part of the habitat fringing the area where most of the chortolirions are found. Currently there are also scattered black wattles elsewhere in the habitat and, if unchecked, this significant stretch of grassland could easily be transformed.

The plants flower and seed well most years, except during severe drought. Some of them, however, even manage to flower in very dry years. The soil around the large sheets of exposed rock remains moist owing to water runoff, even after light rain showers. This moisture ensures that plants in these areas manage to grow and flower while those in drier areas of the same habitat remain semidormant.

Near Delmas in Mpumalanga and in the Carletonville District of Gauteng

The autumn-flowering chortolirions are usually rather sparse, if not rare, in most habitats where they are found. This situation is typified by plants growing around the headwaters of the Bronkhorstspruit near Delmas and the Carletonville hills in western Gauteng, a few hundred kilometres to the west.

The headwaters of the Bronkhorstspruit are surrounded by gently undulating, short grassland. Plants of *Chortolirion angolense* growing here are sparse and scattered. Their rarity is not readily explained since the grassland is burnt most years and livestock tend to be confined to flat areas where the veld grass is taller and the pastures of higher quality. Seed set is good from one year to the next but few seedlings have been found, despite repeated searches.

Most of the hilly areas around Carletonville in western Gauteng belong to various gold mining companies. Agriculture and stock farming have been virtually excluded and the veld is mostly in good condition. These circumstances have made little difference to the densities of autumn-flowering *Chortolirion angolense*. The plants are thinly and widely scattered across the hilly and rocky grassland.

The general rarity of autumn-flowering *Chortolirion angolense* may be associated with the fact that their seed is set late in autumn, from the end of March to early May. This is often a time when conditions are dry and when the early winter veld fires begin. It is suspected that much of the seed that has already been liberated is burnt in winter grass fires. The spring-flowering plants, which are much more abundant, set their seed at the beginning of the main summer rains when conditions for germination are much more favourable.

Growth cycle and the role of fire

The autumn-flowering chortolirions produce their leaves in late November, December and January, a period of frequent summer rainfall. The plants produce leaves and flower amongst living grasses. A fire the previous winter clears dead plant material from the habitat and improves the conditions for the growth and flowering of the plants. This is in contrast to spring-flowering *Chortolirion angolense* for which burning of the grassland is a direct stimulus for flowering; many spring-flowering plants of *C. angolense* flower in a leafless state well before the commencement of the main summer rains.

The peak flowering period is February and early March. Seeds ripen between late March and early May at a time when the veld is drying out and the season of grass fires begins. Seeds are distributed by wind and water runoff after rain. Some are retained in their capsules until June and this may result in the loss of seeds during a grass fire. Seeds have their first chance to germinate the following summer, in October and November, at the beginning of the rainy season.

Status of wild populations

The status of *Chortolirion angolense* depends entirely on what is eventually accepted as the correct placement of the genus and its typification with or without species. If the perspective taken by Craib *et al.* (2004) is correct, then *Chortolirion* can hardly be regarded as vulnerable. Should the distinction between spring- and autumn-flowering populations be accorded some formal recognition, the autumn-flowering plants may qualify for the conservation assessment category 'Vulnerable'.

The situation is entirely different if infraspecific taxa are recognised in *Chortolirion angolense*. In such a case, the plants near Odendaalsrus and Welkom could be considered for the conservation categories 'Vulnerable' or 'Endangered'.

Spring-flowering *Chortolirion angolense* used to be common around Johannesburg and Pretoria. They are still abundant in some parts and seed well year after year despite the densely settled nature of the landscape. The

Right: Specimens of autumn-flowering ***Chortolirion angolense*** amongst tumbled boulders on the hills near Middelburg in Mpumalanga. These plants often have very tall flowering stems, reaching a height of 1 m or more.

remaining populations decrease markedly as the years progress, owing to habitat destruction. It may prove an advantage to identify some areas around Johannesburg and Pretoria where there are large populations of chortolirions that can be set aside as grassland reserves, conserving other plants at the same time.

Vast areas of the interior grassland plateaux of South Africa have been converted to stock farming. Most areas where crops are not produced are devoted to stock farming. Chortolirions are eliminated by crop farming and can exist only in reduced populations where herds of livestock are kept.

Grasslands around large towns and cities often support substantial populations of chortolirions, owing to the limited amount of livestock grazing in the grassland. This situation has been brought about by specific land use patterns but it is hardly representative of chortolirions in most of their grassland habitats.

Cultivation, propagation and horticultural potential

Chortolirion angolense is an unusual plant but it is generally difficult to maintain in cultivation. Plants from the Welkom and Odendaalsrus area, with their ornamental spirally twisted leaves, make attractive container plants.

The bulbs should be planted in well-drained, sandy soil with a good amount of drainage material packed in at the bottom of the pot in which they are grown. Plants tend to grow best in earthenware containers, particularly those with a coarse texture that absorb excess moisture. Chortolirions grow best in positions of morning sun. They should be cultivated in direct sunlight or lightly dappled shade.

The plants should be grown out in the open with plenty of free air circulation and exposed to summer rainfall. They should be watered only during prolonged dry weather. Chortolirions that are grown in greenhouses should be watered regularly from late spring until early autumn, but kept dry in the late autumn and winter. They should be lightly and occasionally watered in the late winter months before the beginning of the growing season in spring.

Chortolirions suffer from root mealybug attack and also from soil compaction when they are grown in containers. The soil should be periodically mulched with finely sieved grass compost and lightly forked over to a depth of less than 1 cm. This practice assists with aeration of the soil.

The plants usually set some seed in cultivation, particularly when they are grown outside in the open. Seeds should be lightly and evenly scattered across the surface of a well-drained germination mix, low in organic content unless pre-treated against fungal attack, and just covered with soil. They should be kept moist before and during germination. Seedlings are prone to attack by damping-off fungi during the entire growing season. They may have to be periodically sprayed with a fungicide. Seedlings should be kept dry in a cool place for the duration of the winter.

Chortolirions are unlikely to become popular in horticulture owing to their specialised cultivation requirements and the difficulty of keeping them alive in the long term.

Cultivation

The methods for cultivating grass aloes and for their propagation from seeds and cuttings have been discussed in detail in the various species accounts in this book. This chapter provides a broader account of the horticulture. It includes the propagation of the various species in nurseries and their appeal to the South African public. A contribution is also made to the debate about the role of indigenous horticulture in the future preservation of rare indigenous South African plants, such as grass aloes.

The chapter is based on interviews with Andrew Hankey, specialist horticulturist at the Walter Sisulu National Botanical Garden in Roodepoort, Linda de Luca, manager of Random Harvest Nursery near Johannesburg and Johan Nieuwoudt, partner in and manager of the Simply Indigenous Nursery situated near the Hartebeespoort Dam in the North-West Province. Both Linda and Johan have an intimate knowledge of propagating aloes, including many grass aloes. Information was also provided by Andrew Hackland, Rachel Saunders and Rodney Saunders, proprietors of Frontier Micropropagation Laboratory in Cape Town. This laboratory has put in place a special programme to micropropagate as many threatened *Aloe* species as possible.

The interview sample was small, consisting of only three people. The interviews were restricted to parties with a detailed knowledge of the propagation and cultivation of many grass aloe species gained through practical experience. A further prerequisite for interviewees in the sample was broad experience of the constraints and possibilities involved in marketing grass aloes.

The research methodology consisted of free interviews around set topics. In the case of the nurseries, the topics ranged from propagation and cultivation techniques to the role indigenous horticulture could play in the future preservation of grass aloes. Questions directed to the horticulturist at the Walter Sisulu National Botanical Garden only revolved around the possible function of indigenous horticulture in *ex situ* conservation.

Frontier Micropropagation Laboratory had micropropagated some of the rarest South African grass aloes and others that generally produced limited amounts of seed in cultivation. These included *Aloe albida, A. kniphofioides, A. linearifolia, A. myriacantha* and *A. vossii*. In all instances, tissue cultures were started from seeds. As many as ten seeds per species were used to provide as much initial genetic variation as possible. It had been found, generally speaking, that grass aloes were easy to multiply in tissue culture since they divided rapidly. Rachel Saunders and Rodney Saunders indicated that they offered micropropagated grass aloe species for sale at the Rare Plant Fair held in Cape Town twice a year. *A. albida, A. linearifolia* and *A. vossii* proved popular amongst the public interested in rare plants. Other aloes such as *A. polyphylla* also sold well.

The techniques used in tissue culture propagation were commercially feasible only when large numbers of plants per species were produced. This implied that outlets for tissue-cultured grass aloes were restricted to wholesale orders placed by the nursery trade. The possibilities for sustaining this level of production were currently very limited in South Africa but might be merited if sufficient numbers of plants could be placed abroad.

Indigenous nurseries in South Africa were often small-scale family businesses aiming to provide limited numbers of a wide range of plants. The managers of both nurseries expressed a preference for propagating as much of their stock as possible from seed. They were reluctant to place orders for tissue-cultured indigenous plants. Reasons advanced concerned an aversion to clonally produced material and the perception that genetic weaknesses were perpetuated via the techniques of tissue culture. One of the nursery managers, however, held the view that tissue culture might prove useful in cases of repeated poor germination in specific grass aloe species.

It has to be remembered that the production of grass aloes in South Africa is at present restricted to small-scale specialist indigenous nurseries. No data could be collected from large-scale general nurseries or garden centres. These enterprises are only in the initial stages of stocking a limited range of indigenous plants. Clonal material may in due course be found to appeal to nurseries stocking mass-produced indigenous plants. If, however, there is no general public interest in aloes, including grass aloes, the existing attitude towards clonally produced material may constrain possibilities.

Tissue-cultured grass aloes were said to have no role in *ex situ* conservation based on the fact that they represented a tiny artificially isolated fraction of the gene pool. If, however, micropropagated grass aloes contributed towards increasing awareness of this group of plants, they had a significant educational function.

Micropropagation techniques usually met with the approval of conservation authorities as they required minimal access to plant material sourced in the wild. There were, however, various constraints within the nursery trade itself with respect to the acceptability of clonal material. All these factors would have to be taken into account at various times in the future if policy decisions were to be based on authentic perspectives.

The methods of propagating and cultivating grass aloes implemented by two nursery managers differ from some of those recommended in the chapters of this book. Their methods are based on different kinds of philosophies, as well as the stock of knowledge accompanying practical nursery experience. Such circumstances lend support to the view that there is no right or wrong way of growing indigenous plants.

One of the two nurseries cultivated aloes and grass aloes in a mixture of river sand, finely sieved, well-rotted compost and seedling mix. A small amount of bone meal was also added to the mixture. Bark and peat moss were avoided as they were found to retain too much moisture, leading to rot.

Seeds were sown from October to January. Later sowings resulted in poor germination during the late summer months. Seeds that did not germinate the previous season sprouted once watering was resumed the following summer.

The germination mixture comprised river sand, fine

silica sand, seedling soil and germination mix. The seeds were lightly covered with river sand to prevent water from moving the seeds around in the trays. They were germinated under cover and irrigated with borehole water. Seeds could not be germinated out in the rain as birds and other small animals excavated the soil in the seed trays.

Seedlings were transplanted when they were three or more months old, when they had produced three or four leaves. They were moved out of the propagation trays as soon as possible to give them plenty of room to grow in spacious containers at an early stage.

The other nursery manager indicated that they used a mixture consisting of silica, sand, bark and compost. The plants thrived in this medium, but problems did arise during periods of continuous rain in summer and some of the plants developed root rot. Aloes, particularly grass aloes, require a good amount of water when in active growth and, although technically succulents, they grow slowly with inadequate water.

The aloes were grown amongst other plants in the nursery, just as they did in nature. In these positions they received some shade from the surrounding plants. They did not trap excessive heat amongst plants displaying a range of sizes and types of foliage.

Seeds were germinated in a commercially obtained mixture consisting of bark, finely milled peat moss and vermiculite. Different amounts of sand were mixed in with each newly purchased load of seedling mix since no batch was ever identical in composition. The seeds were germinated under shade cloth and the germination houses were exposed to natural rainwater, found to be the best trigger for germination and seedling development.

Seeds were sown from late spring to early autumn. It was impossible to get through the volume of seeds that had to be sown by restricting the sowing months to spring and early summer.

Both nurseries experienced some problems with damping off, particularly when seeds were sown too thickly or over-watered during germination. During the germination process, a careful watch was kept on the seed trays for any signs of damping off. One of the nurseries used organic methods to control damping off whenever possible and the other only used an organic fungicide to combat this problem.

One nursery did not experience many problems with pests attacking grass aloes. The plants were grown in an open, well-ventilated environment. When sickly plants were spotted, they were weeded out and burnt. It was felt that nurseries encouraged too many plants to survive. It was better to get rid of the minority of sickly plants than to introduce toxic chemicals into the nursery environment in an attempt to keep them alive. Natural predators were always encouraged in the nursery by virtue of an eco-friendly design, minimising the opportunities for many kinds of pests to establish themselves.

The other nursery experienced some problems with pests, particularly woolly aphids. They were controlled with organic methods whenever practical. This nursery also had an eco-friendly, well-ventilated layout, which helped with pest control.

Both nurseries found that the grass aloe species most in demand were the large, impressive ones such as *Aloe boylei, A. cooperi, A. ecklonis* and *A. kraussii*. There was little interest in the smaller species, particularly those that were difficult to cultivate. Other popular grass aloes were those with unusual growth forms such as *A. woolliana*.

Both had to ensure that their grass aloes reached maturity as quickly as possible. Transplanting the seedlings when they were still young ensured that the aloes were always given conditions for optimum root development.

Both nursery managers were strongly committed to the view that the propagation of grass aloes went a long way towards conserving them in the wild. If plants could be readily obtained from nurseries, they were less likely to be removed from the wild. There was little general knowledge about grass aloes, and the fact that people could obtain them from nurseries contributed markedly to raising levels of awareness about these plants. It was also considered that habitat destruction, habitat degradation and the invasion of the veld by alien plants were continually causing many grass aloe species to decline. As a result of these circumstances, over-collection of grass aloes in the wild would negatively affect many of the species. It was therefore imperative for indigenous nurseries to provide a source of propagated plants.

Both managers were asked if they had formed an opinion about the projected extinction of most South African grass aloes during the course of this century. They held strong beliefs about this matter and their responses were well articulated. It was considered necessary to address the causes for the decline of grass aloes on a prioritised basis. Afforestation of the veld with alien timber trees and invasion of grassland by invasive alien vegetation should be addressed as priorities. In this respect, too much attention was focused on over-regulating nursery production of grass aloes and too little on the factual predicament of the species in their habitats.

It was considered that methods had to be devised that would not only require but also enable landowners to protect endangered flora on their property from alien invasive plants. Considerable progress had been made with recent legislation concerning introduced invasive plants but in practice there remained much to be desired. It was pointed out that grass aloes, amongst other rare South African flora, were continuing to decline at an alarming rate. In this respect it was recognised that conservation departments were under-resourced. Perhaps this and the other factors that had been discussed, required a fresh strategy to conserve grass aloes and other rare South African plants. It was believed that too little attention was given to indigenous horticulture by environmental management agents. Much more progress would be made if it was possible for regular consultation between indigenous nurseries and state-run conservation bodies. Indigenous nurseries had a huge body of knowledge concerning South African indigenous plants.

Before closing this chapter, it is necessary to consider the potential grass aloes may have for creating attractive hybrids. Many species would appeal only to serious aloe collectors. Showy hybrids based on attributes of the different species would be likely to have a wider appeal. Grass aloes have, however, hardly been used for hybridisation programmes in South Africa.

Right: A five and a half year old ***Aloe boylei*** plant with a well-developed root system. It was grown very hard in strong, direct sunlight. The ***Aloe chortolirioides*** seedlings are only two and a half months old. The characteristic thin, grass-like leaves appear only from the second season after germination.

J Condy

Several characters of grass aloes would be useful in hybridisation programmes. Most of them have ornamental foliage. These leaves are furthermore seasonally deciduous, implying that they are not nearly as prone to pests and diseases that affect most other aloes with perennial leaves. Grass aloes often come from very cold areas with regular frosts while they are dormant in the winter. Cold-hardy hybrids could be bred from several grass aloe species. *Aloe modesta* is strongly scented and this species could possibly be used to introduce fragrance into hybrids with showy flowers.

Aloe inconspicua was noted to hybridise accidentally with *Gasteria pulchra*. This intergeneric hybrid displayed the best characters of *A. inconspicua*, namely the long, thin leaves and interesting flowers. Each flower was a delicate shell-pink as opposed to the drab, greenish white flowers borne by *A. inconspicua*.

Aloe verecunda in a rock crevice on a hill in western Gauteng, growing together with *Adromischus umbraticola* subsp. *umbraticola* and *Rhus magalismontana* subsp. *magalismontana*. This forms an attractive natural rock garden.

Grass aloes in a veld garden

The concept of growing grass aloes in a veld garden aims to recreate the natural grassland conditions in which some of the species grow. A veld garden is useful for nurseries that wish to maintain a complete collection of grass aloe species for propagation purposes. This type of garden is likely to play a role in *ex situ* conservation programmes and holds appeal for aloe collectors who wish to grow their plants under natural conditions. The different requirements that various cultivators are likely to have from such a collection are discussed in the text that follows.

A veld garden in which grass aloes are propagated is only possible in areas with naturally occurring grass species where summers are rainy and the winters dry. In South Africa, these conditions are restricted to the summer-rainfall grasslands.

A section of natural short grassland with most grass species growing to a height of 20–40 cm, should be chosen. A piece of land containing sloping and flat ground is ideal since it holds the widest number of habitats in a small area. The aloes should be planted in the grassland under conditions similar to those where they are found in nature. Wherever possible, the plants should be watered by rain, augmented only during prolonged dry spells. The grassland should be burnt towards the end of the winter to stimulate the buds of species that flower after fire.

Nurseries will normally require a good seed harvest from the stock collection, as pure as possible for each species. These objectives are best met by planting grass aloes with different flowering times together and relying on natural pollinators for seed production. A more secure method is to hand-pollinate the plants and to cover them with netting after pollination. The netting has to be sufficiently fine to exclude potential pollinators. Neither of these methods completely excludes the possibilities of cross-pollination and there will inevitably be some hybrids. Each nursery will develop methods best suited to its requirements from these basic guiding principles.

Managers of *ex situ* conservation programmes are unlikely to be interested in seeds from these artificially maintained populations. The possibilities of horticultural contamination during the pollination process are ever present. However, the advantages of this method of cultivation are that some grass aloes grow much better under these conditions than they do in containers. Some of them also sucker readily when the grass is burnt or closely mowed down in late winter.

Aloe collectors sometimes require a group of grass aloes that grow and flower under circumstances that closely imitate those in the wild. These conditions are met in a veld garden and are useful for species with poor records of flowering when cultivated in containers. It has to be borne in mind that the seeds from such collections are of limited use for propagation purposes unless specific arrangements are made for limiting the possibilities of hybridisation to the barest minimum. Natural pollinators are often present in strips of remnant grassland in peri-urban areas. Care has to be taken to ensure that cultivated grass aloes are not grown in veld gardens close to the habitats of naturally occurring grass aloe species. The possibilities of cross-pollination between cultivated and wild plants are enhanced under such circumstances.

A nursery has to establish which grass aloe species are best maintained as mother stock in a veld garden and which fare better when grown in containers and then plan accordingly. Conditions are likely to vary from one area to the next. Some grass aloes always grow best in containers, for example species such as *Aloe soutpansbergensis* and *A. thompsoniae*, which proliferate when grown in hanging baskets under light shade cloth. Other species, such as *A. kniphofioides*, flower well only when conditions approximate those found in their natural environment.

Bulbine favosa frequently grows in association with *Aloe verecunda* on rocky hillsides in western Gauteng and adjacent parts of the North-West Province. The two species usually flower after the first summer rains in November, the bulbines dotting the veld with their yellow blooms interspersed between the flame-orange flowers of *A. verecunda*.

Survival in a changed environment

It is apparent that most of South Africa's grass aloes will enter the rank of extinct species, probably during the course of this century. The timing of these extinctions cannot be accurately projected since the causal factors leading to the demise of the plants are too numerous and variable. Populations of many species have declined and continue to decline at an accelerated rate. This chapter discusses the short- to long-term decline of these plants. It is hoped that the matters raised will contribute to the intellectual debate on preserving these and other grassland plants.

In the world today, science and technology is the dominant ideology with its associated and entrenched behavioural patterns of getting and spending. Life is concerned with material progress. This lifestyle is depleting the resources of the planet and is certainly not sustainable in the long term. The grasslands of eastern South Africa serve predominantly the purposes of producing timber, crops and livestock. As these are such productive areas, it is not surprising that they have been considerably modified by human requirements. This book captures, in its various chapters, the processes of habitat destruction and alteration and the reasons for the widespread decline of the group of plants known as grass aloes. Given the fact that the grassland environment is, and will continue to be, dominated by its use value, the future of grass aloes has to be examined in the light of this situation.

Most grass aloe populations in areas of afforestation with alien timber trees are on the road to extinction in the medium to long term. The levels of human intervention required to manage grass aloes in afforested areas are considerable and likely to be beyond the capacity and resources of the state and the private sector, given the manner in which the use of funds has to be prioritised. At best, certain motivated individuals can apply their skills to the preservation of these plants. The fate of these grass aloes depends entirely on sustaining levels of commitment required for preserving them in successive generations.

In some afforested areas, grass aloes could be used as a resource for *ex situ* conservation. This is a controversial subject but, as the populations of grass aloes in afforested areas continue to decline anyway, this plant resource can be put to another possibly more productive use.

This topic could also be approached from a different angle. The view can be taken that the resources required for preservation of grass aloes in afforested areas are wasted and could be put to other uses. They could, for example, be used to conduct detailed studies on how afforestation has affected ecosystems and the plants living in them. There are certainly many opportunities for post-graduate research in this direction. The body of knowledge gained via research on grass aloes in afforested areas could be used for preserving these and other plants in the few remaining areas that are relatively pristine.

There is still too little detailed information on grass aloes and their pollinators to estimate the minimum requirements for the long-term survival of the different species. Research along these lines, preferably for a master's or doctoral thesis, would provide invaluable information. A major difficulty in this instance would be to derive a model of pristine grasslands and the sum total of potential pollinators in them on which to base altered states in an ecosystem. Such a model would have to be used to assess various stages of alteration in and depletion of grassland ecosystems.

Ex situ conservation is disfavoured by some since the plants are assumed to be removed from the totality of processes in their natural environment. There is also the fear of what is loosely referred to as genetic pollution, brought about by horticultural conditions in *ex situ* conservation initiatives. This perspective has to be examined more closely.

There are few parts of eastern South Africa grassland where grass aloes grow that have been able to retain the spectrum of biological processes that were there once. In many areas, pollinators in particular have been limited to a handful of insects and occasionally sunbirds. This situation was made apparent during two visits to grass aloe populations in peak flower in the spring of 2004. *Aloe chortolirioides* in full flower near Barberton in Mpumalanga was visited only briefly on a few occasions by one possible pollinating insect during three hours of observation under ideal pollination conditions, namely sunny and windless weather. This area is heavily afforested with alien timber trees and well known to exclude a large range of potential pollinators, as a result. *A. woolliana* in flower on the summit of the Steenkampsberg, also in Mpumalanga, was visited repeatedly by about ten different insects during a three- to four-hour observation period. The mountain summit grassland here is relatively pristine. The greater the range of insects that visit grass aloe flowers, the higher the potential that some of them will be active pollinators.

The other major process excluded from areas of afforestation, are burning cycles. In the light of these circumstances it seems correct to take the view that grass aloes are in most cases infrequently found in grassland with biological processes intact, essential for their survival and development and possibly further speciation.

Genetic pollution with respect to grass aloe species is probably only correctly conceptualised in relation to grasslands in an ideal state, with biological processes intact. Boundaries also need to be instated, defining where natural processes remain natural processes, and where genetic pollution begins. It could be argued, for example, that genetic pollution occurs under conditions of alien afforestation. Pollinators, such as there are, visit various species in flower simultaneously, resulting in several hybrids. The chances of this occurring regularly would have been far less likely in natural grassland.

Sound theoretical development is required concerning the principles that can be used to argue in favour of *ex situ* conservation. Perhaps the most pressing question, however, relates to its long-term value, with respect to South Africa's grass aloe species. In this respect, the postgraduate research of Andrew Hankey at the Walter Sisulu National Botanical Garden in Gauteng promises to address a range of these significant questions. Foremost amongst them is the role of *ex situ* conservation in the long-term preservation of grass aloes. It will be necessary to elicit the views of conservation biologists and scientists who support and oppose *ex situ* conservation. The intellectual debates that this research should stimulate may lead to finding common ground necessary for the long-term preservation of these plants.

If the view is taken that *ex situ* conservation of grass aloe species is merited, the practical question remains of how such programmes can be sustained. Grass aloes are a highly specialised group of plants and a succession of dedicated staff with interest and aptitude drawn from many generations will be required to maintain the resource. It may be that the projection of such a future state is not only ambitious but also impossible. It is likely therefore that certain objectives for the management of the *in situ* resource will have to be set in the short to medium term.

Grass aloes in *ex situ* conservation could be held as a resource for introducing some species back into the veld. This will depend not only on the development of further theory around the merits and possibilities of *ex situ* conservation but also on the theoretical developments in conservation biology within the next few decades.

Should consensus be reached that *ex situ* conservation programmes are viable for grass aloes, the stark question remains about where they may be most suitably introduced to the veld. This would have to be in areas of grassland where as many biological processes as possible are still intact. Very little is currently known about this subject and detailed research would have to be carried out on this matter in tandem with any *ex situ* conservation programme.

A final approach to this particular topic is a philosophical contribution. It is apparent that human domination of the planet is going to lead to decades of cataclysmic extinctions of plants and animals. They will disappear as a result of man-made changes to the environment. According to this view, the preservation of what once was, is consequently no longer possible and a new kind of rationality is required to project and manage the future state of the planet. Those who support such a view would argue that this book provides at best a carefully documented and well-illustrated account of specialised plants that are poorly equipped to survive in an altered landscape. The extinction of these plants should be seen only as an opportunity for gaining knowledge about the causal factors that lead to modern plant extinctions. This perspective is very much in keeping with a world view dominated by material progress equated with advances in science and technology.

Habitat degradation and destruction is the primary set of circumstances likely to lead to the extinction of most South African grass aloe species. The impact of horticulture and collecting of these plants is controlled by domestic legislation and the CITES treaty. Many of the grass aloes are Red Data species and access to them or their seeds for horticultural purposes is strictly controlled. From the perspective of domestic conservation, laws are stringently applied since this is one small area where some control over threatened plants can be exercised. Strict application of domestic legislation and the CITES regulations, which are well motivated for the conservation of the plants, does have several negative consequences. The nursery industry tends to shun growing plants for which the requirements for producing them are over-regulated. This is particularly so if the main demand for them is the export market. From the point of view of a nursery, the allocation of time for growing specific plants has to be factored into production costs as with any other business. Producing grass aloes in a nursery is not only difficult, but also tied up with numerous regulations about the export of plants, once they have been successfully produced in the nursery. The consequence is that South African nurseries are unlikely to find it feasible to produce grass aloes for the export market. A future international horticultural demand for these plants may well be met from South African grass aloe species that are already in cultivation abroad and that form a nucleus of propagation stock in countries abroad.

There is a view amongst aloe collectors that grass aloes are difficult to cultivate. It is likely that this perception has prevented the over-collection of many species in the wild. In the case of rarer grass aloe species, collecting them could have negative impacts. The main reason is that many populations of grass aloes have suffered severe habitat loss and degradation. Heavy collecting would only compound the existing problem by leading to the extinction of the better known populations, or those species that are relatively easy to find.

The propagation of grass aloe species in domestic nurseries for local and export horticultural requirements has a role to play in the preservation of these plants. Readily available propagated material removes much of the incentive for lifting grass aloes from the wild. It also adds to the knowledge about techniques for germinating and cultivating the plants. The wider the range of people who produce and cultivate grass aloes, the greater the extent of knowledge about the horticultural requirements of these plants. This knowledge can be of direct benefit to any *ex situ* conservation programmes run by botanical gardens.

One of the most unfortunate problems for the continued existence of many grass aloes in the wild are the projected consequences of global warming. Climatic change on the planet is nothing new but this is the first time such changes have been induced by human activities. Global warming is projected to lead to the extinction of many South African plants as well as to the speciation of others. Those most likely to survive or speciate are plants with large gene pools and relatively intact biological processes. Grass aloes are likely to fare poorly under conditions of increasing aridity and unreliable rainfall. Their populations are too small and a significant number of biological processes are absent from many of the existing populations of most species.

The most practical measures for preserving the grass aloe populations that remain have to involve a number of biological theorists, horticulturists and landowners. Some common ground will also have to be found between the proponents of *in situ* and *ex situ* conservation measures.

In the short term it is certainly necessary to identify grassland areas where the different grass aloe species occur with most of their biological processes intact. The preservation of such areas should become a priority. These populations of plants will require the minimum amount of artificial management for their long-term survival.

References

BAKER, J.G. 1901. Liliaceae 1. In H. Schinz, Beiträge zur Kenntnis der Afrikanischen Flora 13. *Bulletin de l'Herbier Boissier*, Sér. 2: 178–788.

BORNMAN, H. & HARDY, D. 1971. *Aloes of the South African veld*. Voortrekkerpers, Johannesburg.

BURKS, B.D. 1957. A new *Bruchophagus* from a liliaceous plant with a host plant list for the genus (Hymenoptera, Eurytomidae). *Proceedings of the Entomological Society of Washington* 59,6: 273–276.

CRAIB, C., SMITH, G.F., STEYN, E.M.A. & HANKEY, A. 2004. Variation in *Chortolirion angolense* (Asphodelaceae). *Haseltonia* 10: 51–56.

CUMMING, D.M. 2005. *Chortolirion* Berger. Inferred relationships from limited hybridization. *Alsterworthia* 5,2: 2–4.

GLEN, H.F. & HARDY, D.S. 1987. *Aloe cooperi* subsp. *pulchra*. *The Flowering Plants of Africa* 49: Plate 1944.

GLEN, H.F. & HARDY, D.S. 2000. Aloaceae (First part): *Aloe*. *Flora of southern Africa* 5,1,1: 1–167. National Botanical Institute, Pretoria.

HAIGH, L. 2004. Featherstone Kloof. Rehabilitation in a headwater catchment of an Eastern Cape river. *Urban Green File* 8,6 (January February): 32–35. Brooke Patrick Publications, Johannesburg.

HENDERSON, L. 2001. *Alien weeds and invasive plants*. Plant Protection Research Institute Handbook No. 12. Agricultural Research Council, Pretoria.

JEPPE, B. 1969. *South African aloes*. Purnell, Cape Town.

JEPPE, B. 1974. *South African aloes*, 2nd edn, 1st impression. Purnell, Cape Town.

LINDER, H.P. & KURZWEIL, H. 1999. *Orchids of southern Africa*. Balkema, Rotterdam.

PICKER, M., GRIFFITHS, C. & WEAVING, A. 2002. *Field guide to the insects of South Africa*. Struik, Cape Town.

PLOWES, D.C.H. 1986. *Aloe inconspicua*, a new species from Natal. Aloe 23: 32–33.

POLE EVANS, I.B. 1938. *Aloe hlangapies*. *The Flowering Plants of South Africa* 18: Plate 710.

REYNOLDS, G.W. 1936. Notes on two new aloes and one new variety. *Journal of South African Botany* 2: 65–73.

REYNOLDS, G.W. 1942. *Aloe linearifolia*. *The Flowering Plants of South Africa* 22: Plate 849.

REYNOLDS. G.W. 1956. A new aloe from the eastern Transvaal. *Journal of South African Botany* 22: 85–86.

REYNOLDS, G.W. 1966. *The aloes of tropical Africa and Madagascar*. Aloes Book Fund, Mbabane.

REYNOLDS, G.W. 1969. *The aloes of South Africa*. Balkema, Cape Town.

SKEAD, C.J. 1967. *The sunbirds of southern Africa*. Balkema, Cape Town.

SMITH, G.F. 2003. *Aloe craibii* Gideon F.Sm. (Asphodelaceae: Alooideae): a new species of grass aloe from the Barberton Centre of Endemism, Mpumalanga, South Africa. *Bradleya* 21: 25–28.

SMITH, G.F. & CRAIB, C. 2005. *Aloe craibii* and its environment. *Cactus and Succulent Journal* 77,4 : 202–204.

TREUTLEIN, J., SMITH, G.F., VAN WYK, B.-E. & WINK, M. 2003. Phylogenetic relationships in Asphodelaceae (subfamily Alooideae) inferred from chloroplast DNA sequences (*rbcL, matK*) and from genomic fingerprinting (ISSR). *Taxon* 52: 193–207.

VAN WYK, A.E. & SMITH, G.F. 2001. *Regions of floristic endemism in southern Africa. A review with emphasis on succulents*. Umdaus Press, Hatfield.

VAN WYK, B.-E. & SMITH, G. 1996. *Guide to the aloes of South Africa*. Briza Publications, Pretoria.

VAN WYK, B.-E. & SMITH, G. 2003. *Guide to the aloes of South Africa*, 2nd edn, 1st impression. Briza Publications, Pretoria.

VERDOORN, I.C. 1962. *Aloe soutpansbergensis*. *The Flowering Plants of Africa* 35: Plate 1391.

WONGTSCHOWSKI, B. 1990. *Between Woodbush and Wolkberg*, 2nd revised and enlarged edn. Published by B.E.H. Wongtschowski, Haenertsburg, South Africa.

Glossary

Abstract model The sum total of the concept of genus and the concepts of species within it in a plant revision. From the perspective of the production of botanical knowledge, these models derive their authenticity from their ability to display internal consistency amongst the composite elements. Abstract models are often remote from any given or existing conditions in the field. (p. 10)

Alternative revisions of plant genera Different arrangements of the concept of a genus and the concepts of species within it, by plant taxonomists. (p. 12)

Artificial management of grass aloe populations Process concerning active management by human intervention of factors affecting naturally occurring grass aloe populations. It is recommended in grassland areas extensively changed by human activities, such as areas where plantations have become the main form of land use. (pp. 20, 46, 85, 143)

Artificially managed grassland burning regimes Managed burning cycles in grassland, particularly in fire exclusion regions, such as those extensively planted up with timber. (pp. 20, 85)

Autecology That branch of ecology concerned with the study of individual living organisms in their environment rather than many living organisms in one ecosystem. (pp. 1, 2, 10, 12, 20, 36, 62, 65, 111, 118, 123)

Biological processes Life processes of an organism or organisms in one ecosystem. (throughout the text)

Bone meal A fertiliser consisting of ground, pulverised bones used as a slow-release fertiliser for horticultural purposes. (pp. 84, 137)

Burning cycles Times at which fires occur in an ecosystem. (pp. 15, 22, 142)

Bushveld A South African term for vegetation consisting of an upper layer of trees or shrubs and a lower layer of herbaceous plants, often grasses. (p. 7)

Corridors of habitat Thin strips of limited grassland habitat that remain between timber plantations. (pp. 16, 33)

Deep-level mining Mining activities that take place at considerable depth below the surface of the earth. (pp. 1, 18, 116, 118)

Dominant ideology The dominant system of ideas and modes of behaviour prevalent in different eras. (pp. 17, 142)

Ecosystem An interacting system of living organisms and their physical environment which is independent of other groups and self-sustaining, provided it receives radiant energy. (throughout the text)

Endangered Category pertaining to plants in immediate danger of extinction if the causal factors leading to their decline continue to operate. (pp. 102, 134, 138)

Exclusion of fires The active exclusion of fires from areas where they are an integral part of the ecosystem such as in the grasslands of eastern South Africa. (pp. 15, 16, 44)

***Ex situ* conservation** The conservation of an organism, in this case grass aloes, outside its natural environment. (pp. 74, 88, 114, 137, 141, 142, 143)

Feeding episodes Visits paid by sunbird pollinators to grass aloe flowers. The birds usually remain for a period of time feeding on the flowers and then move on in search of other populations of plants in flower, but sometimes return to flowers already visited. (p. 26)

Feeding events Visits by sunbird pollinators to grass aloe flowers in search of nectar usually at the time the flowers are in full nectar flow. Feeding events last as long as the nectar flow, which in itself varies according to the time of the day and weather conditions. (p. 16)

Fire cycles Recurring events when the grassland is burnt, usually during the rainless winter from April to October in South Africa's grasslands. There is usually enough dry grass to permit grass fires each winter. They do not, however, occur every season. Currently, most grass fires are caused by humans whereas in former times lightening strikes were responsible for burning vast areas of grassland. (throughout the text)

Fire ecology That field of study concerned with the role of fire in an ecosystem. South Africa's grassland flora is all part of a fire ecosystem, dependent on fires for growth, flowering and seeding. Exclusion of fire damages grassland ecosystems. (pp. 20, 28)

Fires managed by human intervention Most of the mist belt escarpment habitat frequented by grass aloes is managed by humans to protect extensive timber plantations from destruction by fire. Firebreaks are burnt annually in the same places and other areas of corridor grassland between plantations are burnt infrequently and accidentally. (throughout the text)

Fire prevention areas Areas from which fire is excluded, for example many grassland escarpment regions where timber is produced. This practice effectively removes the grassland from natural fire cycles. (pp. 15, 44, 61, 90, 116)

Fragmented populations Groups of organisms of the same species, grass aloes in this case, isolated from one another by habitat destruction. The main cause of fragmented grass aloe populations are extensive timber plantations. (pp. 57, 61, 71, 85, 127)

Germination mix A commercially available mixture used specifically for the germination of seeds and the development of young plants soon after seeds have germinated. (throughout the text)

Grass fires Fires that occur in grassland, frequently also referred to as veld fires in South Africa. (throughout the text)

Grassland Habitat consisting predominantly of grasses and maintained by fire cycles and low winter temperatures with frequent frost in the high-lying areas. (throughout the text)

Grassland corridors Limited strips of grassland habitat that usually remain after areas have been extensively planted up with commercial timber. They are usually, but not always, left around rocky outcrops, seepage areas or watercourses. (pp. 33, 75)

Habitat alteration Changes to grassland habitats as a result of human intervention. These changes are largely effected by agriculture, mining and commercial timber plantations. (throughout the text)

Habitat destruction Complete ruin of grassland habitats through human intervention such as surface mining activities and the establishment of timber plantations. (throughout the text)

Habitat disturbance Degradation of grassland habitats through changes to the ecosystem such as the invasion by alien vegetation or the frequent incidence of fires. (throughout the text)

Higher levels of abstraction Levels that occur when abstract models, as a knowledge production type in plant taxonomy, are characterised primarily by the internal logic of a classification system (p. 10). See also **Abstract model**.

Highveld A term used for elevated grassland areas maintained by winter fire cycles and frost. (throughout the text)

***In situ* conservation** Conservation of living organisms, in this case of grass aloes, in their natural habitat. (pp. 33, 114, 143)

Intergeneric hybridisation experiments Experiments by which plants belonging to two different genera are crossed to create hybrids. (p. 127)

Invasive alien vegetation Plants introduced from other countries that invade natural areas, degrading and transforming them (throughout the text). See also **Transformation of the grassland**.

Knowledge construction The process of producing knowledge via the use of concepts and research techniques. (pp. 12, 33, 50, 53, 57, 76, 101, 128)

Knowledge-producing communities Groups of people in a specific research field responsible for research leading to the production of various types of knowledge. These communities often share common tertiary education as well as practical training and theoretical orientation. (p. 12)

Knowledge production process The process of producing knowledge via specific orientations towards the use of concepts and theories. (pp. 10, 12, 102, 107)

Kraal manure A South African term for the dung of domestic herbivores usually gathered from kraals (places where livestock is kept), used as a fertiliser for horticultural purposes. (pp. 34, 56, 61, 104)

Land use patterns Types of land use that characterise different geographical areas within specific types of vegetation like grasslands. (pp. 29, 48, 136)

Leaf mould Decomposed foliage that has been broken down to compost. (throughout the text)

Lepidopterous larvae The larvae of butterflies and moths. (p. 27)

Levels of abstraction Layers of theory construction or the use of concepts removed from given or existing conditions. (pp. 10, 112, 127)

Managed burning programmes Specific systems of burning the grasslands managed by humans. The timing of fires as well as their incidence and extent is calculated in advance. This type of burning of the habitat occurs in marked contrast to random grass fires. (throughout the text)

Management of the grassland Specific programmes for managing various ecological processes in grassland such as fire cycles and the control of alien invasive vegetation. (throughout the text)

Marginal stock farming Stock farming that is hardly viable and that is usually practised only for domestic consumption. (p. 18)

Microhabitat Small fragments of a much larger habitat. (pp. 2, 82, 96)

Micropropagate Propagation programmes via the techniques of tissue culture. (p. 137)

Mist belt A geographical region subject to frequent and often prolonged misty weather. These conditions are usually encountered in escarpment areas. (throughout the text)

Model construction The building of intellectual devices that aim to arrange and organise concepts for the purpose of understanding. Taxonomic revisions of plant genera are characterised by most of the elements of model construction. (p. 76)

Model of pristine grasslands An intellectual reconstruction of grassland ecosystems as they conceivably were prior to extensive degradation and destruction by modern humans. (p. 142)

Monotypic genus A genus consisting of a single species. (pp. 127, 128)

Open-cast mining Mining activities that are carried out largely or entirely on the surface of the earth. (pp. 1, 18, 50, 61, 71, 74)

Overburden Material such as rocks and soil not required for the extraction of metals or coal from the surface of the earth. This material is dumped in areas referred to as mine dumps. (p. 18)

Pastoralists Farmers who herd livestock such as cattle, goats and sheep. (pp. 16, 20, 54, 72, 86, 104, 106)

Phenology The study of the times of recurring natural phenomena, especially in relation to climatic conditions, for example the various biological processes (e.g. flowering) in plants that are correlated with the seasons. (p. 127)

Pollination episodes Specific times during which grass aloe pollinators such as sunbird species actively pollinate grass aloe flowers, while feeding on their nectar. (p. 58)

Pollinators The sum total of bird and insect species that pollinate grass aloe flowers. (throughout the text)

Preservation The maintenance of biological processes in grassland and the control of alien invasive vegetation. (throughout the text)

Railway-line reserves Short strips of natural habitat between the railway line and the fenced off adjacent farmland. (throughout the text)

Revisions Regular re-conceptualisation of plant families and genera, resulting in a re-organisation of the taxa classified within them or in a novel conceptualisation of the family or genus itself. (pp. 10, 57, 75, 88, 123)

Riverine grassland corridors Strips of grassland habitat adjacent to rivers in regions extensively planted up with alien timber. (pp. 33, 75)

River sand Finely milled stone resembling the sand deposits that occur in swathes on riverbanks. This gravel is regularly used in horticulture to make up specific types of soil mixtures, particularly for plants that grow best in well-drained mediums. (throughout the text)

Road reserve The area beside a road running parallel to it and usually fenced off from the adjacent farmland. (throughout the text)

Secondary roads Roads with a gravel surface and not sealed off with tarmac. (pp. 9, 29, 96)

Seed chalcids Small wasps the larval stage of which frequently feeds on numerous species of the genus *Aloe*. (pp. 27, 36, 65, 86)

Seedling mix A commercially produced mixture used in horticulture for the growth and development of seedlings. (pp. 34, 98, 137, 138)

Seepage areas Areas that drain off water from the surrounding grassland as well as from sheets of exposed rock. These places are seasonally moist during the rainy summer months. (throughout the text)

Shaft sinking The process of sinking mine shafts for subterranean extraction of gems or metal-bearing ore. (p. 18)

Silica sand Finely milled silica with the consistency of riversand (p. 138). See also **River sand**.

Small-scale pastoralism The keeping and herding of small groups of cattle, goats and sheep for domestic and local village use. (p. 47)

Sociology of botanical knowledge An intellectual investigation of the processes botanists use to construct knowledge about the plant kingdom. This involves the study of changes in the use of concepts, models and paradigm changes (pp. 10, 50, 57). See also **Abstract model** and **Model construction**.

Species concept The system of ideas about how a species should be defined. These take their place within specific frameworks for understanding of the concept of genus. Species concepts are defined by the place they occupy within the construction of botanical knowledge. They may exist within the confines of an abstract model or may refer to plant entities in the field (throughout the text). See also **Abstract model**, **Levels of abstraction** and **Sociology of botanical knowledge**.

Stretching a painting A technique by which buckled paper can be stretched after the painting has been completed by lightly spraying the back of the artwork with clean water, allowing the water to be absorbed by the fibres and then carefully laying the picture on a clean surface, face up, securing it down with gummed tape and leaving it to dry flat. (p. 2)

Stretching art paper A technique by which the paper is gently immersed in a bath of water. It is then held up at two corners to drain off excess water and laid flat on a board. To avoid damage to the fibres, the surface of the paper should not be handled. The paper is then secured on the board with dampened gummed paper and left to dry. Art papers vary in weight, thickness, surface texture and absorbency. Papers with a weight of 180 g/m^2 and less need stretching to prevent buckling. (p. 2)

Stope development The development of mining tunnels under the surface of the earth running adjacent to one bearing rock seams. These tunnels are used for extracting the ore-bearing rock. (p. 18)

Subdisciplines of botany Smaller elements in the discipline of botany, such as autecology. (p. 12)

Sunbirds Bird species of the genera *Nectarinia* and *Anthreptes* that feed on the nectar of many indigenous plants, particularly aloes. (throughout the text)

Sympatric Originating in or occupying the same geographical region, close enough together to be within the range of mutual pollinating vectors. (pp. 39, 91, 101, 102)

Synonymy All that relates to the various scientific names applied to a particular taxon. (pp. 10, 12, 33, 57, 61, 75, 76, 88, 101, 102)

Taxonomic revision The revision by botanists of plant families or genera, usually as a result of innovations in the concepts of family, genus and species. (pp. 10, 75)

Technology and Science as ideology A situation brought about when science and technology occupy the most prominent place in a world view. (p. 17)

Thornveld farm A term used to describe farms, particularly in central KwaZulu-Natal, where various species of *Acacia* trees grow widely scattered across the grassland. (p. 62)

Tissue culture The technique of producing individual plants through the multiplication of various types of plant cells. (pp. 32, 39, 107, 112, 126, 137)

Transformation of the grassland The gradual destruction of various grassland areas by invasive alien vegetation. (throughout the text)

Veld A South African word applied to the countryside, particularly regions of grassland or savanna. (throughout the text)

Veld garden A garden that either preserves existing veld or recreates it in a different setting such as suburbia. These gardens aim either to recreate previous environments or to utilise indigenous plants that might not have occurred naturally in the specific area. (pp. 48, 88, 141)

Veld grass Grass species that occur naturally in the grassland. (throughout the text)

Veld grass compost Compost made from the decomposed stems and leaves of various grass species. (pp. 38, 80, 124)

Vlei A South African term for a low-lying area, often seasonally moist or permanently marshy. (p. 108)

Vulnerable Pertaining to plants that are likely to be classified as endangered should the causal factors that lead to their decline continue to reduce their numbers. (pp. 102, 134)

Well-rotted compost Compost produced at the end of the decomposition process. Such compost is very friable and easy to sieve very finely. (throughout the text)

Index to scientific names of organisms

Names of the grass aloes and the chortolirion treated in this work appear in ***bold italics***, with the main page references in **bold**. All other names are given in *medium italics*.

A

Acacia 13
Acacia dealbata 16, 40, 75
Acacia longifolia 96
Acacia mearnsii 13, 14, 15, 16, 30, 40, 46, 47, 50, 57, 66, 71, 75, 80, 81, 90, 91, 94, 98, 116, 134
Acacia saligna 14, 81, 82, 96
Adromischus umbraticola 116
Adromischus umbraticola subsp. *umbraticola* 140
Albuca 18, 27
Aloe 1, 10, 12, 26, 27, 54, 57, 62, 88, 91, 120, 123, 137
Aloe albida 8, 13, 15, 23, **30–32**, 85, 137
Aloe arborescens 27
Aloe bowiea 10, 12
Aloe boylei 12, 13, 15, 16, 21, 26, 28, **33–35**, 50, 75, 127, 138, 139
Aloe buettneri 12
Aloe bulbillifera 34
Aloe chortolirioides 12, 13, 15, 21, 23, 24, 27, 30, **36–39**, 44, 47, 123, 138, 139, 142
Aloe chortolirioides var. boastii 12, 36
Aloe chortolirioides var. *chortolirioides* 12
Aloe chortolirioides var. *woolliana* 12, 123
Aloe cooperi 7, 8, 12, 13, 23, 33, 34, **39–43**, 66, 76, 138
Aloe cooperi subsp. *cooperi* 12, 39
Aloe cooperi subsp. *pulchra* 12, 39, 40
Aloe craibii 8, 13, 15, 21, 26, 30, **44–46**, 75
Aloe cryptopoda 27
Aloe dominella 6, 12, 13, 29, **47–49**
Aloe ecklonis 12, 14, 16, 18, 19, 22, 23, 26, 28, 29, 33, 34, **50–53**, 57, 58, 60, 61, 75, 76, 116, 138
Aloe fouriei 1, 2, 14, **53–56**, 114, 122
Aloe globuligemma 27
Aloe greatheadii var. *davyana* 27
Aloe hlangapies 2, 7, 12, 13, 14, 15, 16, 23, 24, 29, 50, **57–61**, 116
Aloe inconspicua 14, 19, 21, 24, 26, **62–66**, 91, 127, 140
Aloe integra 6, 13, 14, 15, 21, 23, 40, 44, **66–70**, 72, 116
Aloe kniphofioides 1, 13, 14, 15, 21, 26, 66, **71–74**, 86, 91, 97, 116, 127, 137, 141
Aloe kraussii 13, 16, 26, 28, 33, 50, **75–77**, 138
Aloe linearifolia 8, 9, 13, 14, 18, 44, **78–80**, 114, 137
Aloe lineata 81
Aloe maculata 66
Aloe micracantha 8, 9, 11, 14, 16, **81–84**, 97
Aloe minima 12, 13, 16, 27, 85, 88, 101, 102, 123
Aloe minima* var. *blyderivierensis 12, 13, 27, 85, **88–91**
Aloe minima* var. *minima 12, 27, **85–88**
Aloe modesta 14, 16, 22, **91–94**, 127, 140
Aloe myriacantha 8, 9, 14, 16, 28, 29, 62, **94–98**, 137
Aloe nubigena 1, 2, 8, 9, 12, 13, 23, 44, **98–101**, 107, 111
Aloe parviflora 12, 18, 26, **101–104**, 114
Aloe polyphylla 137
Aloe saundersiae 13, 19, 30, **104–107**
Aloe section *Graminialoe* 10
Aloe section *Leptoaloe* 12
Aloe soutpansbergensis 2, 13, 44, 98, 101, **107–110**, 141
Aloe spicata 27
Aloe suprafoliata 114
Aloe thompsoniae 12, 13, 44, 98, 101, 107, **111–113**, 141
Aloe verecunda 6, 9, 12, 14, 17, 18, 23, 53, 54, 56, 107, **114–118**, 119, 120, 127, 140, 141
Aloe vogtsii 108
Aloe vossii 12, 13, 53, 54, 98, 107, 114, **119–122**, 137
Aloe woolliana 3, 6, 7, 9, 12, 13, 25, 26, 36, 44, **123–126**, 138, 142
Ammocharis coranica 130
Apis mellifera 25, 33, 52, 75, 124
Astylus atromaculatus 52

B

Brachystelma dimorphum subsp. *gratum* 130
Brachystelma modestum 106
Brachystelma nanum 130
Brachystelma pulchellum 102
Brachystelma stenophyllum 130
Brachystelma swazicum 30
Bruchophagus aloineae 27
Bulbine favosa 141

C

Canna indica 80
Catharanthus roseus 80
Ceratotheca triloba 68
Chortolirion 62, 127, 128, 130, 134
Chortolirion angolense 1, 6, 62, **127–136**
Cirsium vulgare 100
Clivia caulescens 100
Crassula 54
Cyrtanthus obliquus 82
Cyrtanthus stenanthus 29

D

Dipcadi 130
Disa venusta 82
Drimia uniflora 75

E

Eucalyptus 14
Eucomis vandermerwei 123
Euphorbia 54
Euphorbia clavaroides var. *truncata* 128
Euphorbia pulvinata 15, 32, 44, 68
Euphorbia restricta 56

F

Frithia minor 134

G
Gasteria pulchra 140
Geranium wakkerstroomianum 21
Gethyllis villosa 127
Greyia radlkoferi 26
Greyia sutherlandii 124

H
Haemanthus montanus 29, 52
Haworthia 127
Haworthia angolensis 62
Haworthia subgenus *Hexangulares* 127

L
Ledebouria 18, 111
Ledebouria dolomiticola 56
Ledebouria zebrina 54
Leucadendron 82

M
Merwilla plumbea 111

N
Nectarinia 15, 25, 58, 68, 71, 72, 116
Nectarinia afra 26
Nectarinia amethystina 15, 26, 33, 46, 75
Nectarinia chalybea 26
Nectarinia famosa 15, 26, 46
Nerine gracilis 29

O
Oldenburgia grandis 9, 11, 81

P
Pelargonium bowkeri 47
Pennisetum clandestinum 50, 78, 81
Pinus patula 44, 90
Pteridium aquilinum 46, 108, 114, 119, 120
Pterodiscus speciosus 130

R
Rhus magalismontana subsp. *magalismontana* 140

S
Selaginella 102
Seriphium plumosum 18
Solanum mauritianum 78, 102, 116
Strelitzia nicolai 9, 14, 78

U
Urginea 130

V
Verbena aristigera 40, 68

X
Xylocopa caffra 38

Z
Zantedeschia jucunda 7, 39, 40

Sponsors Edition

Rolf Bucher

FirstRand Banking Group

Yoshito Iwasa

Tom & Mo McCoy

Julian & Madeline Reed

Collectors Edition

Gustavo & Elizanne Arroyo

Steve Bales

A A Barrell

Toni-Ann & Faye Black

The Brenthurst Library

Paul Brink

Ray Craib

Euclea Books

Liz & Alex Fick

Gariep Plants

Miles Japhet

Caroline Menell

Desmond L Nielsen

Robin M W Parer

Gordon Radcliffe

Reynolds Library, SSSA

Kotie, Arda & Liezl Retief

François Steffens

Malcolm Turner

Unisa Library

Braam & Elsa van Wyk

John Venn

Whitestone Gardens Ltd

Jerry A Windell

Presentation Copies

Gillian Condy

Charles Craib

Murray Ralfe

Braam van Wyk

Standard Edition

A B C Bookshop, Hilton
Graham Ackers
Jeffrey I Anderson
John & Jenny Anderson
Mike & Di Anderson
Susan Andrews
Pere Fraga i Arguimbau
Gustavo & Elizanne Arroyo
Peter J Ashton
Atlas Books

Catherine Babcock
Keith Mitchell Bales
Jonathan B Ball
Marijke & Ken Ball
Dr Marion Bamford
Ruth Bancroft
K R Baragwanath
Bob Barth & Naureen Damude
Auriol Batten
Bellegoni Marcello, Sarzana, Italy
Duke & Kazuko Benadom
Prof Gerhard J Beukes
Bezuidenhoutshoek Farm
Louis A Biagi
Brian J Blaylock
Book League cc
Glenda Booth
Wiebe Bosma
Hans Bosman, Netherlands
Botanical Society of South Africa
Botanischer Garten München
André Botha
Charles & Frances Bower
Sean Boyce
Carolyne Boyce
Nicola Brighton
Dermotte Brown
Dr H J & Mrs M A Brynard
E W & T Buhr
Karen & Gerald Bullen
Shane Burns
John & Sandie Burrows

E Campbell
Robert Guy Carr
Pablo Casado
Jean-Louis Cassel
Jean-Bernard Castillon
Chris, Annie & Saskia Challis
Alex Chanthavong
Michelle Chanthavong
Graham Charles
Clark's Bookshop
The Cockburn Family
Johann & Steph Coetzee
Snooks Cole
Trevor Coleman
Roger & Wendy Colley

Comptoir International du Livre
Gillian Condy
Phil Condy
Margot Connor
Peter Cooke, Boksburg
Keith & Jenny Cowie
Welland & Margie Cowley
Beric & Judy Croome
Neil Crouch
Philip Crous
David M Cumming
Dr A R Currie

Greg & Alice Daniels
J C David
Mrs M Davison
Dave Dawson
Stanley Dawson
Martin Dawson-Brown
Ivone de Figueiredo & Jan van Loggerenberg
Linda de Luca
Paul de Luca
Dr & Mrs C Deacon
Eileen Dean
Christian Defferrard
C J & M J Dekker
David D Desper, Jr
Mark Detterman
N J Devenish
Sara Dewar
Giancarlo Discepoli
Geraldine & Bill Dittrich
Kenneth Doble
Maria Dodds
Dr Philip E Downs
Francois & Ronelle du Randt
Mary Parisi & Ed Dunne
Bruce & Andi Durow
Dennis P Durow
John & Sue Duthie

Urs Eggli
Chester Emery

Joan Faiola
Phil Favell
Luuts Feenstra, The Netherlands
Nora Flavin
Kirby W Fong
Phillip Alan Foster
Eugene & Lalie Fourie
Luigi Franco
Robin Frandsen
Russel, Bonnie & Gabriella Friedman
Georg & Hannes Fritz

Ian Gaigher
Peter V Gammarano, Jr
Mervyn & Allison Gans

Itai Gartushka
Gerrit Germishuizen
David & Gaby Gess
Fred & Sarah Gess
Howard & Monica Gie
Alice Gilbert
Thomas Glavich
Sean Gildenhuys
Goldfields Library, Florida, Gauteng
José M Gracia Santacreu, Calpe, Spain
Markus Graf
Great Books, Umhlanga
Dave Green
Anita & Jürgen Greve
G, K, A & J Grieve
Dr Alberto Grossi
Dr Gottfried Gutte

Susanne A Haffner
Norbert Hahn
A C J Hall
Andrew & Michelle Hankey
Anna Hargreaves
Mark & Bridget Harman
Revd Diana Harpham
John F Harrison
Harry Molteno Library, SANBI
Dr Hayashi M
Geoffrey A Hedgecock
Hermannsburg Nature Club
Lem & Pat Higgs, California
Norbert Hoffmann
Malcolm Holloway
Horley Family
D J Human
Maureen Patricia Hutton

Siegfried Janssen
Dr Jan Martin Jecminek
R G Jeffery
Jo, Ben, Rori & Gabi
Paul A M Joubert
Alain Jubault

Sidney H Kahn
Bernd & Waltraud Keller
Brian Kemble
Judd Kirkel
Nicholas Anthony Klapwijk
Chris F Kleiss
Ronell & Arrie Klopper
Koeltz Scientific Books
Erich Kuschke
Teuns Kuun

Dave Lambie
Suzanne Lategan
V I Latti
John Lavranos

M Le Fur Camille
Maddy Lehmann
Alison Lettenga
John Levengood
B G Lindeque
Henk & André Loots
Lawrence M Loucka
Abraham J Lourens
Johan Louter
Adriaan Hendrik Louw
Peter & Margo Louw

Megan Mac Nair
P MacQuilkan, Managa Hills, Magalies
Drs Daniel & Susan Mahr
Harry Chi-King Mak
C W Malan
Michael Marais
A E March
Daniel Masson, Botanist, Geneva
Erica McCoy
Gavin McDonald
Derick McKenzie
Cameron & Rhoda McMaster
Douglas McMurtry
Richard Meir
Adv Abri Meiring
Victor Millard
John B Miller
Jeff Miller
Dorothy Minors
Missouri Botanical Garden, USA
Buddy & Jenny Mockford
Tiaan & Jessica Moolman
A P Mortimer
Moullec Jean-Michel
René & Elmarié Mulder
Dirk Muller

Len Newton
Geoff, Lynne & Douglas Nichols
John & Penny Nicholson
Desmond L Nielsen
Johann & Leigh Nieuwoudt

Andy & Pat Ochse
Michael Oliver
Connall Oosterbroek

Giacomo Paonessa
Parceval (Pty) Ltd
Robin M W Parer
Herby Peel
Robyn Pells
Dr R M Pelteret
Lauriel Irene Petersen
Nanette Pienaar
Elsa Pooley
Ken Price

Dina Prinsloo

Melvyn Quan

John Rae
Rainbow Gardens Bookshop
Random Harvest Nursery
David & Nicky Rattray
Mark Read
Norbert Rebmann, France
Red Pepper Books
Gwen Rigby
Brenda Robson
Jan Romijn
Ingrid Roos

San Francisco Succulent Society
Rod & Rachel Saunders
Georg Schelinski, Keltern
David M Schultz
Michael & Karoline Schurr
Dr Shirley Sherwood
Gaye Simmons
Simply Indigenous Nursery
Drs Joe & Jill Skowno
Jenny & Charles Smith
Michael H Solomon
Peggy M Spaete
Sarel & Madeleine Spies
John Stathoulis
Lucas & Estelle Steenkamp
Dr Norbert Steffen
David & Isolde Stegmann
Lizelle Stone
Roger & Wilna Stones
Susan Strauss
E Ströh
Dawie & Ebeth Strydom
D J Supthut
Andrew, Vanessa & Ryan Sutherland
M B E Sweet

Audrey Tang
Greg Tarr
Leo Thamm
Patrick & Linda Thomas
Frank R Thorold
Hans-Jürgen Thorwarth
Hugh Tregarthen
Sarah Treherne
Roy Trendler
Dave & Jo Trickett

B uit de Bosch

Erika van den Berg
Dick & Liz van der Jagt
P P van Deventer
Christo & Ida van der Merwe
Bill van der Riet
Jill & David van der Schyff
Bertus van Eeden
Maxwell van Os
Jacob van Pareen
C van Wijk
Susan, Stephan & Vellies Veldsman
Patrick Vernemmen
Venessa Viney
Dr & Mrs P Vorster

R D Wall
Ann Wanless
Dr I S Watt
Robert Wellens
Robert Welsh
Ilse & Stephan Wentzel
Johan Wentzel
Whitestone Gardens Ltd
Wilmart Madeleine
Els & Nigel Wildman
Derric H Wilson
Beverley Wilson
Pieter Winter
Wisley Library RHS Gardens
Frank Wolpert
Rob & Janet Wood
Albina & Mike Wortley